Springer-Science+Business Media, B.V.

Dear Reader

We would very much appreciate receiving your suggestions and criticisms for the *Plant Analysis Manual.* They will prove to be most helpful during our preparations for future supplements.

Would you please answer the questions listed below, and send your comments with any further suggestions you may have to *Ir. Ad C. Plaizier* at the abovementioned address.

Thank you for your assistance!

AF769052

Ir. Ad C. Plaizier
Publisher

PLANT ANALYSIS MANUAL

1. What errors have you found? (list page numbers and describe mistakes)
2. What protocols do you find to be confusing or lacking in detail? (list chapter numbers and page numbers and describe problems)
3. What protocols do you feel should be replaced in future supplements with newer (better) methods?
4. What new topics or other material would you like to see included in future supplements?

Please print or type your answers in the space below and continue overleaf.

Name: Date:

Address:

PLANT ANALYSIS MANUAL

PLANT ANALYSIS MANUAL

Edited by

I. WALINGA

J.J. VAN DER LEE

V.J.G. HOUBA

W. VAN VARK

I. NOVOZAMSKY

Springer-Science+Business Media, B.V.

Library of Congress Catalog Card Number

Plant analysis manual / edited by I. Walinga ... [et al.].
p. cm.

1. Plants--Analysis. I. Walinga, I.
QK865.P58 1995
581.19'2--dc20 94-35634

ISBN 978-94-011-7651-4 ISBN 978-94-011-0203-2 (eBook)
DOI 10.1007/978-94-011-0203-2

Printed on acid-free paper

Contents

This Manual has been issued in 1989 as a completely revised and extended edition of the former "Methods of Analysis of Plant Material" by Van Schouwenburg & Walinga, originally meant for use within the M.Sc. course in Soil Science and Water Management.

With this edition of the Manual, it has been reorganized and put into a new, more handsome format. The Manual is now strictly arranged according to the methods of digestion/extraction, and all corresponding determination procedures are now placed in the *same* chapter. Thus, each determination procedure is tailored to the conditions of the digest/extract one is working with. We hope that this will avoid any confusion which may arise when combining procedures. Besides, we trust that the new, loose-leaf format will be more practical in the laboratory setting.

The authors accept no responsibility whatsoever for any harm caused by the application (right or wrong) of these procedures.

Any questions, comments, suggestions for improvements etc. should be adressed to

Dr. V.J.G. Houba
Wageningen Agricultural University
Department of Soil Science and Plant Nutrition
P.O. Box 8005
NL-6700 EC Wageningen
The Netherlands

Introduction

0.1 ORGANIZATION OF THIS MANUAL

The Plant Analysis Manual is intended for the practizing (agricultural) chemist when analyzing plant material. Therefore, the present Manual only contains *ready-to-hand procedures* without any comment.

The determination procedures described here are only for *inorganic* components, most frequently elements. Besides, most procedures are designed to give a *total* content value of the element under consideration, regardless of the chemical structure in which it occurs in the plant.

Sampling and conservation of plant material are beyond the scope of this manual; some general remarks are, however, compiled in section 0.3. It is anyhow assumed that the material, which is handed in for analysis, is *representative* for the purpose of the customer.

In the field of plant analysis there is a confusing variety of methods and procedures, both for digestions and determinations; moreover, in many cases the digestion and the subsequent determination are interrelated. For example, a separate digestion is needed for trace elements in order to obtain determinable concentrations. Contrary to the former issue of this Manual, we have now chosen for a design in which the digestion/extraction procedure is described in one chapter together with all determination procedures that may be carried out on that particular digest/extract. All the necessary extra information (like standardizations) appear in *appendices*. As a consequence, several determination procedures are described two or three times, however, each based on a particular digestion or extraction method.

Two types of determination procedures are described: *manual* and *automated*. The *manual* procedures are mainly used in research laboratories, whereas the *automated* procedures are more frequently applied in routine laboratories. Both types of determinations can be used freely, provided that appropriate equipment is available.

For determination of the elements, mainly *spectrometric* techniques are used here. Depending on the kind of element and the expected concentration level, the following methods are applied: flame atomic absorption spectrophotometry (flame AAS), graphite furnace atomic absorption spectrophotometry (furnace AAS), flame emission spectrometry (flame AES), plasma emission spectrometry (ICP-ES), absorption photometry (colorimetry). Besides, potentiometric and coulometric titration will be encountered.

In several cases, more than one method is described to determine a component. This provides a reference, as well as an *alternative* in case of instrumental or analytical problems.

Each procedure is *fully written down*, without reference to likewise procedures, since it is very inconvenient to be forced to riffle through the pages while performing the analysis. Often, however, one and the same determination procedure can be applied to different digests/extracts. In these cases, the procedure is fully written down for each of the relevant types of digestion/extraction, because the standard series should be made with different acids so as to match the acid type and strength of the sample digests/extracts in question.

A great many of the methods in this Manual have been *checked* in our laboratory, which is reflected by specific remarks.

Literature references are only sparingly mentioned; in principle only if these provide specific extra information. This may be the first article on the subject in question, but it may be a recent (review) paper as well.

In general, no warnings are given with respect to *safety*. It is assumed that every analyst and laboratory technician realizes that all chemicals are unsafe, although to a different extent. Exceptions are only made in the case of a particularly toxic or harmful substance, e.g., KCN. For the rest it is assumed that all pipetting is done with the help of pipette fillers, that safety goggles are worn when performing digestions, that work is done in a fume hood when harmful fumes may be released, etc. There is very much literature on the subject of laboratory safety; it is strongly recommended to consult such safety manuals and to observe the safety rules given therein.

No special references are made with respect to the impact of these procedures on the *environment*. The user should realize, however, that the waste from a laboratory may be very detrimental for the environment, even in small amounts. This is especially true for transition metals (e.g., Cr from chromic acid), unsaturated hydrocarbons (e.g., benzene), and halogen containing organic compounds (e.g., chloroform, tetra). Users are strongly recommended to take appropriate measures in order to minimize these effects.

0.2 TECHNICAL REMARKS

In applying the procedures described in this manual, one should realize that there is a number of underlying assumptions, agreements and conventions. For clarity and brevity, these are summarized below, so that there is no need to repeat them with every procedure.

- The common simple *laboratory equipment and glassware* (hot plates, beakers etc.) is supposed to be available. If special equipment, glassware or plastic ware is needed, it will be mentioned explicitly.
- In many cases, the classical volumetric pipette can be substituted by *plunger-type pipettes* or by *dispensers*, provided that these are calibrated. This is in particular convenient for dispensing concentrated acids, including sulphuric acid. For work where high precision is required, the use of a volumetric pipette is still recommended.
- All reagents are assumed to be of analytical grade, unless otherwise indicated.
- It is assumed that all glassware has been *cleaned with 4 M nitric acid.*
- When *water* is used, either demineralized or distilled water is meant. Double-distilled, "demi-dest" or otherwise extra purified water is seldom needed, and will only then be prescribed.
- In *weighing* a prescribed amount of sample or reagent, it is ordinary custom that the actually weighed mass may deviate ± 10 %, since such small differences can be corrected arithmetically. We would like to emphasize, however, that it may be convenient to weigh out exactly the prescribed amount; although this will take some more time, it might compensate the effort of extra calculations and the inherent risk of calculating errors.
- Procedures are pointed at consumption of *minimum* amounts of reagents. Thus, a sentence may read "dissolve 2 g in 50 ml" when this is the normal amount for one run. You may prepare, however, 10 g in 250 ml, or any multiple, if you intend to do more runs. With automated methods, however, all reagents are given per litre, since the consumption simply depends on the number of determinations per day.
- The number of significant digits is an indication of the required *precision.* Thus, 1000 ml is far more precise than 1 litre. By applying this convention, there is no need to prescribe the required glassware, so that the user can select appropriate equipment according to his own findings. As exception to the rule, the use of volumetric flasks is explicitly prescribed for stock solutions. In practice, volumetric flasks are very convenient to handle, e.g., for mixing, so that these may be used even though their intrinsic precision is not needed.
- The *atomic weights* used for calculation are taken from an official IUPAC table. The relevant values have been compiled in Appendix A-1. Molecular weights have consequently been calculated from these values, and eventually rounded off.
- *SI units* have been used throughout this manual; in a separate Appendix (A-2) attention is paid to their use. Within this system there is a preference for the use of basic units; thus, the density of water should be given in kg m^{-3}. For better readability, however, we have chosen for the use of g/ml if this does not cause ambiguity. We have also chosen for the consequent use of *molarity* in stead of normality. We realize that this might lead

to difficulties in the calculation routine. It should therefore be pointed out, that – although we prefer to stick to the SI system – everyone is free to apply normality for own use.

- For calculation of the results, a *condensed formula* is given. In this way, anyone can set up his/her own way of calculating, whereas the final result can be checked by the given formula.
- To express the content of a compound in plant material, the analytical chemist prefers the use of moles, but the agricultural chemist uses (milli)grams for ease of interpretation. For this manual we have chosen to adhere to the generally accepted usage of *mmol/kg* and *mg/kg* for macro elements, and *μg/kg* for micro elements.
- It may be mentioned here, that any laboratory should take measures to improve and maintain the *quality of its analytical results*. In particular, the use of certified reference samples and an active membership of interlaboratory trials are appropriate means for this purpose.

0.3 SAMPLING AND PRETREATMENT

Although the present Manual is not intended to describe sampling and pretreatment procedures for plant material, these stages are sufficiently important to mention here the main considerations and rules.

The *sample* must be representative for the whole lot of plant material. This is a crucial point for the customer, since the chemical composition may vary strongly, dependent on plant part (leaf, stem, fruit), physiological age and growth conditions. Besides, it is necessary to collect as much plant material as possible in order to minimize variations due to heterogeneity.

Pretreatment of the sample involves drying and grinding, preceded by washing if the fresh plant material is (likely to be) polluted. Soil, dust and salts from irrigation water are the usual contaminants; these may be washed out by tap water, 0.1 M hydrochloric acid or 1% detergent solution, followed by rinsing with demineralized water. The internal concentration of major nutrients will not significantly be affected by this treatment, if the washing does not take more than 30 s.

The plant sample is normally dried at 70 °C in a well-ventilated drying oven during 24 h. The material is then finely ground, in order to obtain a homogeneous sample from which representative subsamples can simply be taken. As a rule of thumb, the milled plant material should pass a 1-mm sieve when less than 1 gram is to be weighed out. Both drying and milling should be carried out with equipment that does not release elements for which the samples are to be analyzed. One should realize, for instance, that a mill may contaminate the sample with Al, Cd, Cu, Fe, Pb and possibly other heavy metals, depending on its composition.

The dried and milled samples should be stored in a cool and dry place in tightly stoppered flasks or in sealed polythene bags, protected against direct sunlight. During storage, the plant material may attract moisture so that the drying procedure must be repeated just before weighing out a sample for analysis.

The analytical results are often referred to "oven-dry" material, which means dried at 105 °C. For comparability, therefore, the moisture content should be determined by drying at 105 °C and taking the difference with the 70 °C dried sample. The drying at 105 °C should be done, however, with a separate sample, since this operation may change its chemical composition.

REFERENCES

M.G. Keizer, V.J.G. Houba and Th.M. Lexmond. Sampling of soil and plant material for chemical analysis. Syllabus Department of Soil Science and Plant Nutrition, Agricultural University, Wageningen, The Netherlands (1984). (In Dutch)

C. Sonneveld and P.A. van Dijk. The effectiveness of some washing procedures in the removal of contaminants from plant tissue samples of glasshouse crops. Commun. Soil Sci. Plant Anal. *13* (1982) 487–496.

1 Digestion in **tubes** with H_2SO_4-salicylic acid-H_2O_2 and selenium and determination of Ca, K, Mg, N, Na, P, Zn

1.0 DIGESTION IN **TUBES** WITH H_2SO_4-SALICYLIC ACID-H_2O_2 AND SELENIUM

Field of application

This digestion is in particular suited for routine work on large series of plant samples and automated determinations. It can be applied for the determination of Ca, K, Mg, Mn, N-total, Na, P, and Zn in plant material.

Principle

The larger part of organic matter is oxidized by hydrogen peroxide at relatively low temperature. After decomposition of the excess H_2O_2 and evaporation of water, the digestion is completed by concentrated sulphuric acid at elevated temperature under the influence of Se as a catalyst.

Remarks:
- Salicylic acid is added to prevent loss of nitrate.
- A precipitate of $CaSO_4$ may be formed when cooling after completing the digestion; it will dissolve in 18–20 hours after the addition of water. Therefore, Ca can be measured only after this period.

Apparatus

- Aluminium heating block with holes for digestion tubes.
- Metal weighing funnels with long spouts (see Fig. 1)
- Digestion tubes, 100 ml, with narrowed neck (see Fig. 2).

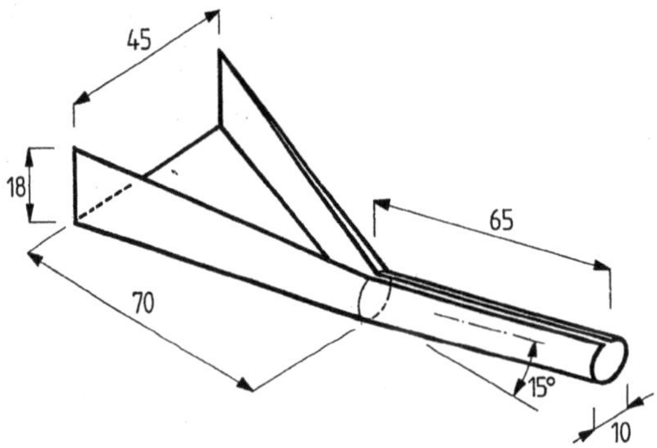

Fig. 1. Weighing funnel.

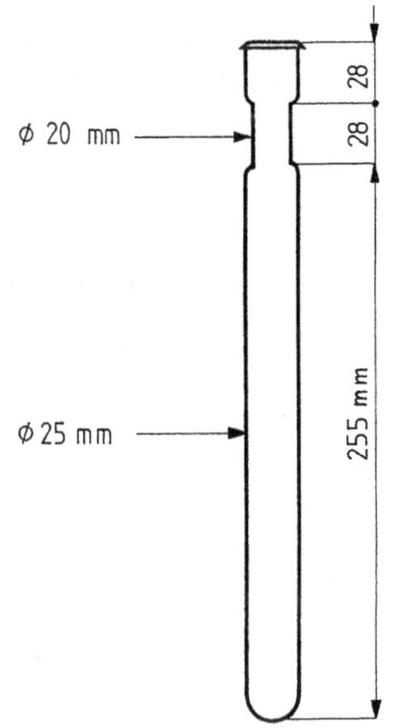

Fig. 2. Digestion tube.

Remarks:
- Dried plant material may easily stick to glass when the relative humidity of the air is low. These home-made weighing funnels (stainless steel or aluminium) do not show this effect.
- These weighing funnels have been designed with an extra long spout, so that the plant material is released *below* the narrowed neck.
- The dimensions of digestion tubes may be different from those indicated in Fig.2, as long as they fit exactly into the holes of the heating block used and have a length of at least 15 cm.
- In the authors' laboratory, a series consists of 30 tubes: 24 samples, 2 blanks, 1 plant sample with known low concentration (in duplicate) and 1 with known high concentration (also in duplicate). These known samples serve as internal quality control.

Reagents

(1) Sulphuric acid, 96% (w/w), $c(H_2SO_4)$ = 18 mol/l (d = 1.84 g/ml).
(2) Hydrogen peroxide 30% (w/w).

Remark:
- Hydrogen peroxide must be of analytical quality; a lower grade may be stabilized with EDTA, phosphate, or other interfering compounds.

(3) Selenium, powder.
(4) Salicylic acid, powder.
(5) Sulphuric acid-selenium mixture. Dissolve 3.5 g of selenium (3) in 1 litre of sulphuric acid (1) by heating to about 300 °C, while covering the beaker with a watch glass. The originally black colour of the suspension turns via green/blue into clear light-yellow. The entire process takes 3–4 hours.
(6) Digestion mixture. Dissolve 7.2 g of salicylic acid (4) in 100 ml of the sulphuric acid-selenium mixture (5). This digestion mixture should not be stored for more than 48 h.

Remark:
- In the authors' routine laboratory, where this digestion is applied every day, the plant material is dried again at 70 °C just before weighing; at that moment its moisture content is only 1–2%, so that concentrated sulphuric acid may be used for the digestion. When the drying is not repeated, the plant sample may contain up to 10% moisture. The use of concentrated sulphuric acid then causes a raise in temperature which results in loss of nitrate. In that case, diluted sulphuric acid should be used as described in Chapter 2 (digestion in flasks).

Procedure

Weigh, to the nearest 0.001 g, approximately 0.3 g of the dried plant material sample in a metal weighing funnel and transfer the sample to a digestion tube. Take care that all the plant material comes below the narrow part of the tube. Add 2.5 ml of the digestion mixture (6) and swirl carefully until all the plant material is moistened. Allow to stand for at least 2 h. Prepare also two blank digestions.

Place the tube in the heating block and heat at 100 °C for at least 2 h. Remove the tube from the block, allow to cool, and add successively three 1-ml aliquots of hydrogen peroxide (2). Mix carefully, but thoroughly after each addition. WARNING: the reaction is violent. Wait until the reaction with hydrogen peroxide has ceased (approximately 10 s) before adding the next portion. Place the tube again in the preheated block and heat at 330 °C. The digestion is considered complete when the digests have turned colourless or light-yellow; this usually takes about 2 h. Remove the tube from the block and cool to room temperature. Add 48.3 ml of water and mix. Allow to stand overnight. Mix again, transfer the digest to a test tube and let settle.

Remarks:

- After moistening the sample with digestion mixture (6), at least 2 h are needed for the nitro-salicylic acid compounds to be formed; in the authors' laboratory this period is often taken overnight.
- At least 2 h at 100 °C are necessary to obtain complete reduction of the nitro-salicylic acid compounds.
- The boiling point of the digestion mixture (6) lies just above 330 °C; when the block temperature is adjusted at 330 °C, the temperature in the tube will be approximately 300 °C.
- The volume of digest is remarkably constant: 2.1 ± 0.1 ml for the samples, and 2.5 ml for the blanks. For that reason, it is appropriate to add a fixed volume of 48.3 ml water to the digest by means of a dispenser. According to our experience, any volume marks on tubes lead to fairly large errors, since the tubes are wide and the marks are often inaccurately placed.
- When Zn is to be determined, no rubber stoppers should be used.
- SiO_2 will dissolve gradually and may then interfere in the determinations. Thus, the determinations should be done soon after digestion, otherwise the digest has to be filtered.
- The sulphate concentration in the blank digest is 0.8 mol/l.

REFERENCES

I. Novozamsky, V.J.G. Houba, R. van Eck and W. van Vark. A novel digestion technique for multi-element plant analysis. Commun. Soil Sci. Plant Anal. *14* (1983) 239–249.

1.1.1 DETERMINATION OF **CALCIUM** BY FLAME AES

Field of application

This determination is intended to be carried out on digest 1.0 (H_2SO_4-salicylic acid-H_2O_2-Se, in tubes).

With calcium concentrations higher than 1 mol/kg dry material, the solubility product of $CaSO_4$ can be surpassed in the digest; a smaller sample has to be weighed out then.

Principle

The sample is vaporized in an air-acetylene flame; calcium compounds are atomized and the CaOH molecules thus formed emit radiation of which the intensity is measured at a wavelength of 622 nm.

Remarks:

- Although Ca measurements are also possible at 423 nm and 554 nm, the spectral interferences at 622 nm are smaller for elements present in plant material, whereas larger interferences at this wavelength only come from elements that hardly or not occur in plant material.
- When using the cooler air-propane flame the emission signal is 30–40% lower, and there will be more interference.

Apparatus

- Flame emission spectrometer.

Reagents

(1) Hydrochloric acid, c(HCl) = 4 mol/l. Add 340 ml of concentrated hydrochloric acid (36%) to about 400 ml water and make up to 1 litre.
(2) Stock solution, Ca concentration 5000 mg/l. Weigh out 12.486 g of calcium carbonate, $CaCO_3$, pretreated according to Appendix A-4. Transfer it to a 1000-ml volumetric flask with the help of about 150 ml water. Add 63 ml of 4 M hydrochloric acid (1) and boil to expel CO_2. If calcium carbonate particles remain visible, add 5 ml of 4 M hydrochloric acid (1) extra. Allow to cool and make up to volume with water.

Standard series

Pipette 0–2.00–4.00–6.00–8.00–10.00 ml of the stock solution (2) into 100-ml volumetric flasks, to which already 50 ml water and 4.5 ml of concentrated sulphuric acid (96%) per flask have been added in advance. Make up to the mark with water.

Dilute this standard series 1 + 9 (v/v) with water; this should be done together with the diluting of the sample digests. This diluted standard series has Ca concentrations of 0–10–20–30–40–50 mg/l.

Procedure

Dilute all digests 1 + 9 (v/v) with water.

Measure the emission intensity of the diluted digests, the diluted blank digests and the diluted standard series at a wavelength of 622 nm, using an air-acetylene flame. Plot a calibration curve and read the calcium concentrations.

Remarks:

- The calibration curve should be linear, but shows a slight bending towards the x-axis. This is due to ionisation, which could be counteracted by addition of a cesium salt. The bending is, however, so small that Cs addition is not necessary.
- In determination 1.1.2 a releasing agent is added to prevent calcium atoms from being caught into poorly dissociating compounds, in particular oxides. Such an addition is not necessary here, because calcium is measured at an oxide band.
- When an air-propane flame is used, there may be interference from Fe and Al, which is counteracted by addition of magnesium chloride. This causes, however, such a higher salt level that the nebulizer may readily be clogged; therefore, water should be nebulized very frequently.
- Instead of indicating a wavelength, a simple emission spectrometer ("flame photometer") may only be supplied with a so-called Ca filter. Make sure that this is an interference filter, since glass filters are not selective enough.

Calculation

The calcium content of the dried plant material, expressed in mmol/kg Ca, is calculated by

$$0.2495 \times (a - b) \times \frac{V}{w}$$

in which
a = concentration of calcium in the diluted sample digest, in mg/l;
b = concentration of calcium in the diluted blank digest, in mg/l;
V = total volume of digest at the end of the digestion procedure, in ml;
w = weight of plant material sample, in g.

1.1.2 DETERMINATION OF **CALCIUM** BY FLAME AAS

Field of application

This determination is intended to be carried out on digest 1.0 (H_2SO_4-salicylic acid-H_2O_2-Se, in tubes).

With calcium concentrations higher than 1 mol/kg dry material, the solubility product of $CaSO_4$ can be surpassed in the digest; a smaller sample has to be weighed out then.

Principle

The sample is vaporized in an air-acetylene flame; calcium compounds are atomized and the calcium atoms thus formed absorb radiation from a hollow-cathode lamp. The absorbance is measured at a wavelength of 422.7 nm.

Remark:

- Since Ca atoms are easily captured in the flame into poorly dissociating compounds (like oxides), a releasing agent like lanthanum must be added.

Apparatus

- Atomic absorption spectrophotometer.

Reagents

(1) Hydrochloric acid, c(HCl) = 4 mol/l. Add 340 ml of concentrated hydrochloric acid (36%) to about 400 ml water and make up to 1 litre.
(2) Lanthanum solution, La concentration 1 g/l. Dissolve 3.12 g of lanthanum nitrate hexahydrate, $La(NO_3)_3 \cdot 6H_2O$, in some water and make up to 1 litre with water.
(3) Stock solution, Ca concentration 1000 mg/l. Weigh out 2.497 g of $CaCO_3$

(pretreated according to App. A-4) and transfer it with 100–150 ml water to a 1-litre volumetric flask. Add 13 ml of 4 M hydrochloric acid (1) and boil to expel CO_2. If calcium carbonate particles remain visible, add 1 ml of 4 M hydrochloric acid (1) extra. Allow to cool and make up to volume with water.

Standard series

Pipette 0–2.00–4.00–6.00–8.00–10.00 ml of the stock solution (3) into 100-ml volumetric flasks, to which already 50 ml water and 4.5 ml of concentrated sulphuric acid (96%) per flask have been added in advance. Make up to the mark with water. This standard series has Ca concentrations of 0–20–40–60–80–100 mg/l.

Procedure

Pipette 0.25 ml of the sample digests, the blank digests and the standard series into test tubes. Add 4.75 ml of lanthanum solution (2). Nebulize each measuring solution into a yellow (reducing) air-acetylene flame and measure the absorbance at a wavelength of 422.7 nm. Plot a calibration curve and read the calcium concentrations.

Remark:
- The calibration curve should be linear by virtue of the releasing action of lanthanum.

Calculation

The calcium content of the dried plant material, expressed in mmol/kg Ca, is calculated by

$$0.0250 \times (a - b) \times \frac{V}{w}$$

in which
a = concentration of calcium in the sample digest, in mg/l;
b = concentration of calcium in the blank digest, in mg/l;
V = total volume of digest at the end of the digestion procedure, in ml;
w = weight of plant material sample, in g.

1.2.1 DETERMINATION OF **POTASSIUM** BY FLAME AES

Field of application

This determination is intended to be carried out on digest 1.0 (H_2SO_4-salicylic acid-H_2O_2-Se, in tubes).

The determination may be done in combination with Na (1.6.1).

An automated procedure is described in section 1.2.2.

Principle

The sample is nebulized into an air-propane flame, where it is vaporized; potassium compounds are atomized and the potassium atoms thus formed emit radiation of which the intensity is measured at a wavelength of 766.5 nm.

Remark:

- To prevent ionisation interferences, a cesium salt is added, to act as an ionisation buffer.

Apparatus

- Flame emission spectrometer.

Reagents

(1) Cesium solution, Cs concentration 0.87 g/l. Dissolve 1.1 g of cesium chloride, CsCl, "pro analysi", in 1 litre of water.

(2) Stock solution, K concentration 4000 mg/l. Dissolve 7.627 g of potassium chloride, KCl, in some water in a 1000-ml volumetric flask and make up to the mark with water.

Remarks:

- The cesium chloride should be of the highest analytical grade ("pro analysi"), because a lower quality (e.g. "reinst" = "most pure") contains much more Na.
- Frequently, K determinations go along with Na determinations. In that case, a mixed stock solution may be prepared with 4000 mg/l K and 1250 mg/l Na (see 1.6.1 – reagent 2).

Standard series

Pipette 0–2.00–4.00–6.00–8.00–10.00–12.00 ml of the stock solution (2) into 100-ml volumetric flasks, to which already 50 ml water and 4.5 ml of concentrated sulphuric acid (96%) per flask have been added in advance. Make up to the mark with water.

Dilute this standard series 1 + 9 (v/v) with the cesium solution (1); this should be done together with the diluting of the sample digests. This diluted standard series has K concentrations of 0–8–16–24–32–40–48 mg/l.

Procedure

Dilute all digests 1 + 9 (v/v) with the cesium solution (1).

Measure the emission intensity of the diluted sample digests, the diluted blank digests and the diluted standard series at a wavelength of 766.5 nm, using an air-propane flame. Plot a calibration curve and read the potassium concentrations.

Remarks:

- The calibration curve should be nearly linear.
- An air-acetylene flame can be used also, since the Cs concentration is high enough to counteract the greater tendency to ionisation.
- Instead of indicating a wavelength, a simple emission spectrometer ("flame photometer") may only be supplied with a so-called K filter. Make sure that this is an interference filter, since glass filters are not selective enough.
- When using a filter in the spectrometer, the emission from K at the wavelength of 769.9 nm is also measured. This enhances the sensitivity of the K measurement.

Calculation

The potassium content of the dried plant material, expressed in mmol/kg K, is calculated by

$$0.256 \times (a - b) \times \frac{V}{w}$$

in which

a = concentration of potassium in the diluted sample digest, in mg/l;
b = concentration of potassium in the diluted blank digest, in mg/l;
V = total volume of digest at the end of the digestion procedure, in ml;
w = weight of plant material sample, in g.

1.2.2 DETERMINATION OF **POTASSIUM** BY FLAME AES (AUTOMATED, BY FLOW ANALYZER)

Field of application

This determination is intended to be carried out on digest 1.0 (H_2SO_4-salicylic acid-H_2O_2-Se, in tubes).

The potassium content of the dried plant material should be higher than 30 mmol/kg.

The determination may be done in combination with the automated determination of N (1.5.3), Na (1.6.2), and P (1.7.2) by flow analyzer.

The manual determination is described in section 1.2.1.

Principle

In a continuous-flow analysis (CFA) system, the sample is mixed with an internal standard (Li) and then vaporized in an air-propane flame. Potassium compounds are atomized; the potassium atoms thus formed emit radiation of which the intensity is measured at a wavelength of 766.5 nm.

Apparatus

- Continuous-flow analysis system (sampler, pump, K/Na unit, flame photometer). In the authors' laboratory, a Skalar Segmented Flow Analyzer is used, but any CFA system will do.

  Note: Depending on the flame photometer used, potassium can be determined really simultaneously with sodium in this automated procedure.
- Digital printer.

Reagents

(1) Wash fluid. Only demineralized water without any wetting agent should be used as wash fluid, in order to prevent contamination of the flame photometer by potassium.

(2) Internal standard. Dissolve 12.7 g of lithium chloride, LiCl, in 1000 ml water. Add 7.5 ml of this solution to about 800 ml water; add also 12.5 ml of glycerol, $C_3H_8O_3$, and make up to 1000 ml. Prepare just before use.

Remark:
- The internal standard may vary according to the flame photometer used. In the authors' laboratory, therefore, a "Radiometer solution" – supplied by the manufacturer of the flame photometer – is used as internal standard.

(3) Mixed stock solution, N concentration 2500 mg/l, P concentration 500 mg/l, Na concentration 1250 mg/l and K concentration 4000 mg/l. Weigh out 11.793 g of ammonium sulphate, $(NH_4)_2SO_4$, 2.1978 g of potassium dihydrogen phosphate, KH_2PO_4, 3.178 g of sodium chloride, NaCl, and 6.424 g of potassium chloride, KCl. Dissolve these in bidest in a 1000-ml volumetric flask and make up to the mark with bidest.

Remarks
- In the authors' laboratory, the mixed stock solution (3) is used because these four elements are determined in one run. Should the phosphate be left out, then more potassium chloride is needed for the potassium standard series.
- The mixed stock solution (3) and the standard series are made with bidest in favour of the P determination; for the potassium/sodium determination it is not necessary.

Standard series

Pipette into 100-ml volumetric flasks, that contain already about 50 ml water, 4.5 ml of concentrated sulphuric acid (96%). Allow to cool. Then add 0–4.00–8.00–12.00 ml of the mixed stock solution (3) and dilute to volume.

Dilute this standard series 1 + 9 (v/v) in polycarbonate test tubes with water; this should be done together with the diluting of the digests. The diluted standard series has K concentrations of 0–16–32–48 mg/l (and Na concentrations of 0–5–10–15 mg/l).

Procedure

Dilute all digests 1 + 9 (v/v) with water in polycarbonate test tubes.

Make up a measuring series by placing standards and samples in the automatic sampler in the following sequence:

1–3	fourth term of the standard series (highest standard)
4	first term of the standard series (zero standard)
5	second term of the standard series
6	third term of the standard series
7	fourth term of the standard series (highest standard)
8	first term of the standard series (zero standard)
9	blank digestion
10–48	rest of the digestion series (i.e., 34 samples, two standard samples "low", two standard samples "high" and one more blank digestion)
49–50	fourth term of the standard series (highest standard).

Put the continuous-flow system into operation according to the scheme given herewith; rinse the system and the flame photometer during 5 minutes with demi water and next for 30 minutes with reagents; then put the sampler into operation.

Measure the K emission intensity ratio with respect to Li. Print the K-Li emission intensity ratios. Plot a calibration curve and calculate the K concentrations.

After finishing the measurements (at the end of the day), flush the flame photometer for 15 min with water.

Remarks:

- It is advantageous to use a printer instead of a recorder. The moment of printing should, however, be adjusted carefully so as to coincide with the top of the peak. Once adjusted, the tube length from the 10-turn coil to the flame photometer must remain constant.
- The standard at position 3 in the sampler should give the same value as those at positions 49 and 50.
- The zero standard (position 8) may give too high a result because of influence of the highest standard measured just before. Therefore, this observation is rejected.
- In the authors' laboratory, the sample stream is actually split up directly after the pump into three parallel streams: one for K/Na, one for N, one for P.

Calculation

The potassium content of the dried plant material, expressed in mmol/kg K, is calculated by

$$0.256 \times (a - b) \times \frac{V}{W}$$

in which

a = concentration of potassium in the diluted sample digest, in mg/l;

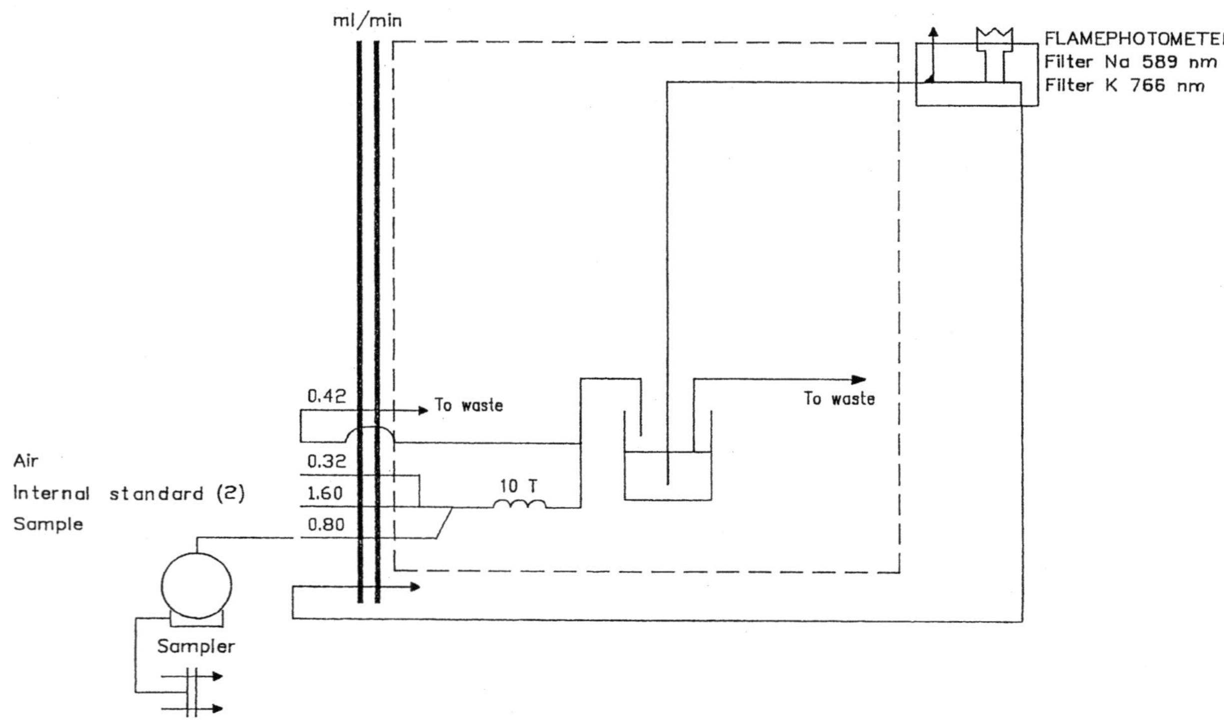

Flow Diagram K/Na-determination (1.2.2.).

b = concentration of potassium in the diluted blank digest, in mg/l;
V = total volume of digest at the end of the digestion procedure, in ml;
w = weight of plant material sample, in g.

1.3 DETERMINATION OF **MAGNESIUM** BY FLAME AAS (AUTOMATED)

Field of application

This determination is intended to be carried out on digest 1.0 (H_2SO_4-salicylic acid-H_2O_2-Se, in tubes).

The magnesium content of the dried plant material should be higher than 10 mmol/kg.

Principle

The sample is nebulized into an air-acetylene flame, where it is vaporized; magnesium compounds are atomized and the magnesium atoms thus formed absorb radiation from a hollow-cathode lamp. The absorbance is measured at a wavelength of 285.2 nm.

Remark:
- Since Mg atoms are easily captured in the flame into poorly dissociating compounds, in particular with phosphate and aluminium, a releasing agent like lanthanum must be added.

Apparatus

- Diluter/dispenser, allowing high dilution factors at low analyte volumes while maintaining high precision.
- Atomic absorption spectrophotometer, provided with a printer and possibly with an automatic sampler, and allowing readings in concentration units with curved calibration lines.

Reagents

(1) Lanthanum solution, La concentration 1 g/l. Dissolve 3.12 g of lanthanum nitrate hexahydrate, $La(NO_3)_3 \cdot 6H_2O$, in some water and make up to 1 litre.
(2) Standard solution, Mg concentration 500 mg/l. Dissolve 5.0705 g of magnesium sulphate heptahydrate, $MgSO_4 \cdot 7H_2O$, standardized according to App. A-20, in some water in a 1000-ml volumetric flask and make up to volume.

Standard series

Pipette 0–2.00–4.00–6.00–8.00 ml of the standard solution (2) into 100-ml volumetric flasks, to which already 50 ml water and 4.5 ml of concentrated sulphuric acid (96%) per flask have been added in advance. Make up to the mark. This standard series has Mg concentrations of 0–10–20–30–40 mg/l.

Procedure

(a) Dilution

Using the diluter/dispenser, pipette 0.12 ml of the sample digests, the blank digests and the standard series into test tubes, and add 4.68 ml of the lanthanum solution (1). Note, that the diluted standard series has Mg concentrations of 0–0.25–0.50–0.75–1.00 mg/l.

(b) Calibration of the AA spectrophotometer

Set the wavelength of the atomic absorption spectrophotometer at 285.2 nm; set the other operating parameters (lamp current, gas flow, observation height etc.) at their – previously elaborated – appropriate values for magnesium determination. Light the flame and adjust the acetylene flow and/or the air flow such that the flame is just blue (stoichiometric). Nebulize the diluted standard series and enter the nominal concentration values; the spectrophotometer is now calibrated in concentration units.

(c) Measurement

Nebulize the diluted digests and measure the magnesium concentrations at a wavelength of 285.2 nm and furthermore the same measuring conditions at which the spectrophotometer was calibrated. After intervals of 10 samples, measure the diluted zero standard and the diluted second standard (0.5 mg/l) again.

If the read concentration differs from the entered concentration, adjust the zero reading and then adjust the sensitivity until the nominal concentration is displayed again. Next, measure the foregoing ten samples again, beginning with the last sample, until the readings differ no more.

Remarks:

- In the authors' laboratory, no automatic sampler is used for this determination. If the use of such a sampler is considered, one should be aware that this implies tedious arithmetic corrections afterwards if the sensitivity has changed.
- No interferences were observed after spiking the digests with 200 mg/l Al, 600 mg/l PO_4 or 200 mg/l SO_4.

Calculation

The magnesium content of the dried plant material, expressed in mmol/kg Mg, is calculated by

$$1.648 \times (a - b) \times \frac{V}{w}$$

in which

a = concentration of magnesium in the diluted sample digest, in mg/l;
b = concentration of magnesium in the diluted blank digest, in mg/l;
V = total volume of digest at the end of the digestion procedure, in ml;
w = weight of plant material sample, in g.

1.4 DETERMINATION OF MANGANESE BY FLAME AAS

Field of application

This determination is intended to be carried out on digest 1.0 (H_2SO_4-salicylic acid-H_2O_2-Se, in tubes).

The determination may be carried out in combination with Zn (1.8).

Principle

The sample is nebulized into an air-acetylene flame, where it is vaporized; manganese compounds are atomized and the manganese atoms thus formed absorb radiation from a hollow-cathode lamp. Lanthanum is added to prevent condensed phase interferences. The absorbance is measured at a wavelength of 279.5 nm.

Apparatus

- Atomic absorption spectrophotometer.

Reagents

(1) Sulphuric acid, $c(H_2SO_4) = 0.8$ mol/l. Add carefully, while swirling, 45 ml of concentrated sulphuric acid (96%) to about 400 ml water. Allow to cool and make up to 1 litre.
(2) Lanthanum solution, La concentration 16.5 g/l. Dissolve 12.86 g of lanthanum nitrate hexahydrate, $La(NO_3)_3 \cdot 6H_2O$ in water and make up to 250 ml.
(3) Stock solution, Mn concentration 500 mg/l. Dissolve 1.438 g of potassium permanganate, $KMnO_4$, (standardized according to Appendix A-21), in about 200 ml of water in a beaker. Add carefully, while stirring, 45 ml of concentrated sulphuric acid (96%). Reduce the permanganate with a few drops of hydrogen peroxide (30%) and boil to remove the excess of H_2O_2. Transfer the contents of the beaker quantitatively to a 1000-ml volumetric flask and make up to the mark.
(4) Standard solution, Mn concentration 50 mg/l. Pipette 50.00 ml of the

stock solution (5) into a 500-ml volumetric flask and make up to volume with 0.8 M sulphuric acid (1).

Remark:

- Frequently, Mn determinations go along with Zn determinations. In that case, a mixed standard solution may be prepared with 25 mg/l Mn and 20 mg/l Zn (see 1.8 – reagent 3).

Standard series

Pipette 0–1.00–2.00–4.00–6.00–8.00–10.00 ml of the standard solution (4) into 100-ml volumetric flasks and make up to volume with 0.8 M sulphuric acid (1). This standard series has Mn concentrations of 0–0.5–1–2–3–4–5 mg/l.

Procedure

Pipette 5.00 ml of the sample digests, the blank digests and the standard series into test tubes. Add 0.50 ml of lanthanum solution (2) and mix. Nebulize all these solutions into a blue (oxidizing) air-acetylene flame and measure the absorbance at 279.5 nm. Plot a calibration curve and read the manganese concentrations.

Remarks:

- The calibration curve is slightly bent towards the x-axis, so that calculation by means of linear regression is not allowed.
- In a 0.08 M sulphuric acid medium, no interferences were found after spiking the digests with various salts, so as to reach final concentrations of 80 mg/l Al, 500 mg/l Ca, 80 mg/l Fe, 800 mg/l Mg or 16 mg/l Si. However, a concentration of 1000 mg/l Ca results in a precipitate of $CaSO_4$ if the solution has been left standing overnight.

Calculation

The manganese content of the dried plant material, expressed in mg/kg Mn, is calculated by

$$(a - b) \times \frac{V}{w}$$

in which
a = concentration of manganese in the sample digest, in mg/l;
b = concentration of manganese in the blank digest, in mg/l;
V = total volume of digest at the end of the digestion procedure, in ml;
w = weight of plant material sample, in g.

1.5.1 DETERMINATION OF **NITROGEN** BY COLORIMETRY

Field of application

This determination is intended to be carried out on digest 1.0 (H_2SO_4-salicylic acid-H_2O_2-Se, in tubes).

If, for any reason, this determination cannot be carried out, then determination 1.5.2 may be done.

An automated procedure is described in section 1.5.3.

Principle

The determination is based on the Berthelot reaction, in which a phenol derivative (here: salicylate) forms a coloured compound in the presence of ammonia and hypochlorite. In alkaline medium, the indophenol thus formed has a green-blue colour, of which the absorbance is measured at a wavelength of 660 nm. This is a measure for the concentration of ammonium, formed by the nitrogen compounds in the sample.

Remarks:

- To prevent precipitation of hydroxides, EDTA should be added prior to raising the pH.
- The pH should be adjusted with a buffer solution, since the digests contain variable amounts of acid.
- Sodium nitroprusside is a catalyst for the Berthelot reaction.

Apparatus

- Colorimeter.

Reagents

(1) Sodium hydroxide solution, c(NaOH) = 10 mol/l. Dissolve 200 g of sodium hydroxide, NaOH, in about 400 ml water; allow to cool and make up to 500 ml.

(2) Salicylate solution. Dissolve 110 g of salicylic acid, $C_7H_6O_3$, in 105 ml of sodium hydroxide solution (1) and make up directly with water to 250 ml. Prepare just before use.

(3) Buffer solution pH 12.3. Dissolve 26.70 g of disodium hydrogen phosphate dihydrate, $Na_2HPO_4 \cdot 2H_2O$, in some water in a 2-litre volumetric flask. Add 10 ml of the sodium hydroxide solution (1) and dilute to volume with water. Measure the pH and adjust if necessary.

(4) EDTA solution. Dissolve 4 g of disodium ethylene diamine tetra acetate dihydrate, $Na_2EDTA \cdot 2H_2O$, in 100 ml water.

(5) Hypochlorite solution. A stock solution, containing approximately 1 M sodium hypochlorite in 0.1 M NaOH, should be purchased commercially. Dilute 20 ml of this stock solution with water to 100 ml. Prepare fresh daily.

(6) Nitroprusside solution. Dissolve 50 mg of sodium nitroprusside dihydrate, $Na_2[Fe(CN)_5NO] \cdot 2H_2O$, in 100 ml water. Prepare just before use.
(7) Mixed reagent I. Mix 50 ml of the salicylate solution (2) with 100 ml of the nitroprusside solution (6) and 5 ml of the EDTA solution (4).
(8) Mixed reagent II. Mix 200 ml of the buffer solution (3) with 50 ml of the hypochlorite solution (5).
(9) Stock solution, N concentration 2500 mg/l. Dissolve 11.793 g of ammonium sulphate, $(NH_4)_2SO_4$, in water in a 1000-ml volumetric flask and make up to volume with water.

Remarks:
- The final solution of sodium hypochlorite (5) should contain 0.7% ± 0.1% of active chlorine. This will be true for solutions purchased from a supplier of chemicals; if, however, the hypochlorite was bought as bleach in a supermarket, a check on its concentration is recommended.
- The mixed stock solution of 1.5.3 can also be used.

Standard series

Pipette into 100-ml volumetric flasks, that contain already about 50 ml water, 4.5 ml of concentrated sulphuric acid (96%). Mix and let cool. Then add 0–1.00–2.00–3.00–4.00–5.00–6.00 ml of the stock solution (9) and dilute to volume with water.

Dilute this standard series 1 + 9 (v/v) with water; this should be done together with the diluting of the digests. The diluted standard series has N concentrations of 0–2.5–5 –7.5–10–12.5–15 mg/l.

Procedure

Dilute all digests 1 + 9 (v/v) with water.

Pipette 0.20 ml of the diluted sample digests, the diluted blank digests and the diluted standard series into test tubes. Add 3.0 ml of mixed reagent I (7) and mix. Then add 5.0 ml of mixed reagent II (8) and mix. Allow to stand for at least 2 h. Measure the absorbance in a 1-cm cuvette at a wavelength of 660 nm. Plot a calibration curve, and read the N concentrations.

Remarks:
- The highest standard (15 mg/l N) should give an absorbance reading of approximately 0.68 units.
- In 2 hours the blue colour reaches its maximum intensity; it is stable for at least 10 h.
- The test tubes must be used exclusively for this N determination. They must first be cleaned by taking a blank determination; thereafter they should be cleaned only with water.

Calculation

The nitrogen content of the dried plant material, expressed in mmol/kg N, is calculated by

$$0.714 \times (a - b) \times \frac{V}{w}$$

in which

a = concentration of nitrogen in the diluted sample digest, in mg/l;
b = concentration of N in the diluted blank digest, in mg/l;
V = total volume of digest at the end of the digestion procedure, in ml;
w = weight of plant material sample, in g.

REFERENCES

I. Novozamsky, R. van Eck, J.Ch. van Schouwenburg and I. Walinga. Total nitrogen determination in plant material by means of the indophenol blue method. Neth.J.agric.Sci. *22* (1974) 3–5.

1.5.2 DETERMINATION OF **NITROGEN** BY TITRATION AFTER DISTILLATION

Field of application

This determination is intended to be carried out on digest 1.0 (H_2SO_4-salicylic acid-H_2O_2-Se, in tubes).

Although this determination gives satisfactory results, it is somewhat outdated; therefore, determination 1.5.1 is preferred.

Principle

Ammonia, originating from digestion of nitrogen compounds, is distilled over into boric acid solution. The borate thus formed is titrated with hydrogen iodate to the colour change of an acid-base indicator.

Remarks:

- Since $H(IO_3)_2^-$ is a strong acid, the ammonium borate formed by the distillation is titrated to boric acid; thus, the consumption of $H(IO_3)_2^-$ is directly proportional to the amount of ammonia. As a consequence, neither the volume nor the concentration of the boric acid solution used to capture the ammonia needs to be known accurately.
- Sulphuric acid is often used for the titration. Potassium hydrogen iodate, however, has the advantage of being a primary standard. Yet, the procedure should be checked (see below).
- Although any acid-base indicator can be used, it is known that the sharpest end points are obtained with mixed indicators. Here, a mixture of bromocresol green and methyl red is used.

Apparatus

- Steam generator, electrically heated (see Fig. 3).
- Distillation unit ("Markham still"), micro or semi-micro design (see Fig. 4).
- Piston burette.

Remarks:

- This steam generator is protected against overloading; the production of steam is adjusted automatically according to the consumption of steam. In the authors' laboratory, a Büchi DG-1500 steam generator is used (manufactured by Büchi, CH-9230 Flawil, Switzerland).
- This still has the evaporator vessel enclosed in a combined steam jacket and trap: an arrangement which is compact and accelerates distillation. In the authors' laboratory, a Quickfit 46 MC distillation unit is used (manufactured by Corning Ltd., Laboratory Division, Stone, Staffordshire ST15 0BG, England).

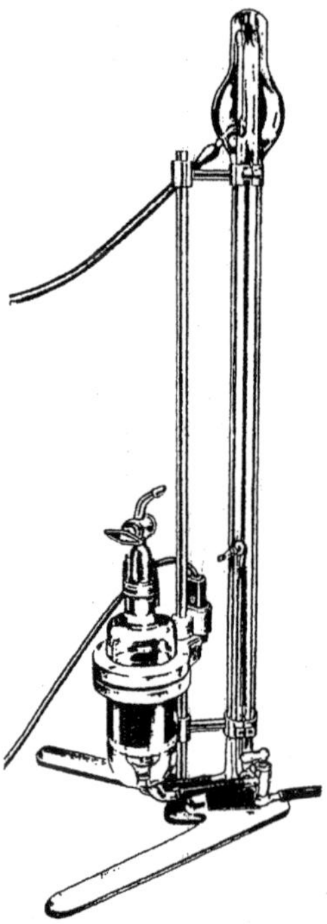

Fig. 3. Steam generator.

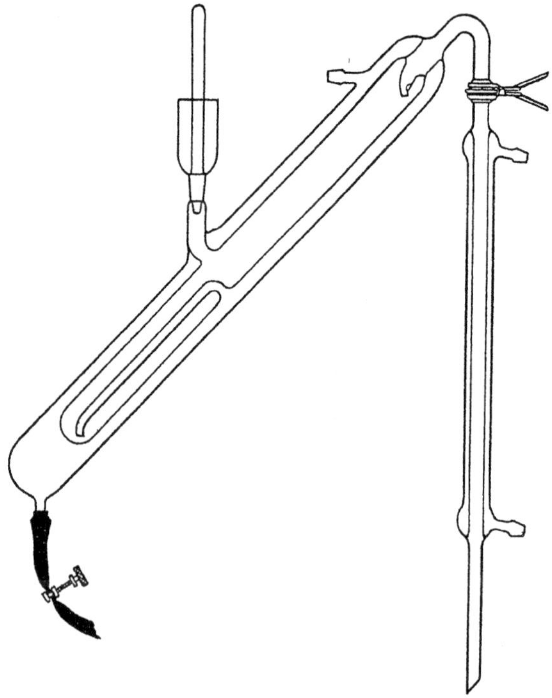

Fig. 4. Distillation unit.

Reagents

(1) Sodium hydroxide solution, c(NaOH) = 12.5 mol/l. Dissolve carefully, in portions, 250 g of sodium hydroxide, NaOH, in about 400 ml water. Allow to cool and make up to 500 ml with water.

(2) Boric acid solution. Dissolve 10 g of boric acid, H_3BO_3, in 1 litre of water.

(3) Indicator solution. Dissolve 0.15 g of bromocresol green and 0.1 g of methyl red in 200 ml of ethanol 96% (v/v).

(4) Hydrogen iodate solution, $c(H(IO_3)_2^-)$ = 0.01 mol/l. Dissolve 3.90 g of potassium hydrogen iodate, $KH(IO_3)_2$, in 1000 ml water.

(5) Standard solution, $c(NH_3)$ = 0.0100 mol/l. Dissolve 0.6607 g of ammonium sulphate, $(NH_4)_2SO_4$, in about 400 ml water in a 1000-ml

volumetric flask. Add 45 ml of concentrated sulphuric acid (96%), allow to cool and make up to volume.

Standard series

A full standard series is not necessary, but the procedure should be checked by means of standard solution (5) and a zero standard solution, for two reasons. Firstly, the actual procedure may give a result that deviates from the theoretically expected relationship between the amounts of hydrogen periodate and ammonia. Secondly, the optimum distillation time should be established for the apparatus to be used.

Procedure

(a) Standard solution

Pipette 5.00 ml of standard solution (5) into the distillation unit. Add 2.5 ml of sodium hydroxide solution (1) and rinse with some water to a total volume of about 20 ml. Heat the liquid with steam; when the first drop of distillate appears, continue distilling during n minutes (n = 3, 4, 5 or 6), while leading the distillate into 10 ml of boric acid solution (2), to which 6 drops of indicator solution (3) have been added. Take care that the end of the condenser is immersed in the boric acid solution to prevent any loss of ammonia.

Allow the inner side of the condenser tip to be washed by the last portions of distillate by means of lowering – 0.5 min before the distillation is finished – the receiver with boric acid until the end of the condenser is out of the solution.

Titrate the boric acid solution with hydrogen iodate solution (4) until the colour changes from green to rose.

Repeat this procedure with different distillation times. Choose the optimum time and use this when distilling the samples.

Repeat the procedure with a zero standard solution.

Calculate the exact titer of the hydrogen iodate solution, using the results found with the optimum distillation time and correcting for the zero standard, by

$$t_{HIO3} = 0.0500 \times \frac{1}{(V_s - V_z)}$$

in which

t_{HIO3} = titer (concentration) of the hydrogen iodate solution (4), in mol/l;

V_s = volume of hydrogen iodate solution consumed for titrating the standard solution, in ml;

V_z = volume of hydrogen iodate solution consumed for titrating the zero standard solution, in ml.

(b) Sample solution

Pipette 5.00 ml (containing at most 0.1 mmol NH_3) of digest into the distillation unit. Add 2.0 ml of sodium hydroxide solution (1) and rinse with some water to a total volume of about 20 ml. Heat the liquid with steam and distil during the previously established optimum time, in the way described at (a).

Titrate the boric acid solution with standardized hydrogen iodate solution (5) until the colour changes from green to rose.

Repeat the procedure with a blank digest.

Calculation

The total nitrogen content of the dried plant material, expressed in mmol/kg N, is calculated by

$$200 \times t \times (a - b) \times \frac{V}{w}$$

in which

t = concentration of the hydrogen iodate solution, in mol/l;
a = volume of hydrogen iodate solution consumed by the sample digest, in ml;
b = volume of hydrogen iodate solution consumed by the blank digest, in ml;
V = total volume of digest at the end of the digestion procedure, in ml;
w = weight of plant material sample, in g.

1.5.3 DETERMINATION OF **NITROGEN** BY COLORIMETRY (AUTOMATED, BY FLOW ANALYZER)

Field of application

This determination is intended to be carried out on digest 1.0 (H_2SO_4-salicylic acid-H_2O_2-Se, in tubes).

The nitrogen content of the dried plant material should be higher than 150 mmol/kg.

The determination may be done in combination with the automated determination of K (1.2.2), Na (1.6.2) and P (1.7.2) by flow analyzer.

The manual determination is described in section 1.5.1.

Principle

The determination is based on the Berthelot reaction, in which a phenol derivative (here: salicylate) forms a coloured compound in the presence of ammonia and hypochlorite. In alkaline medium, the indophenol thus formed has a green-blue colour, of which the absorbance is measured at a wavelength of 660 nm. This is a measure for the concentration of ammonium, formed by the nitrogen compounds in the sample.

Apparatus

- Continuous-flow analysis system (sampler, pump, nitrogen unit, photometer, recorder). In the authors' laboratory, a Skalar Segmented Flow Analyzer is used, but any CFA system will do.

Reagents

(1) Sampler wash fluid. Bidest without any wetting agent.
(2) Wash fluid 1. Dilute 21 ml of concentrated hydrochloric acid (36%) with water to 250 ml.
(3) Wetting agent Aerosol 22, to be purchased from Sigma Chemical Co., product number A 9753.
(4) Wash fluid 2. Add 2 ml of wetting agent (3) to 1 litre of bidest.
Note: This wash fluid is only meant for rinsing the CFA system.
(5) Wetting agent Brij 35, 30% w/v solution, to be purchased from Sigma Chemical Co., product number 430AG-6.
Note: This wetting agent is only meant for addition to reagents (6), (7) and (8).
(6) Saline diluent. Dissolve 9 g of sodium chloride, NaCl, in 1 litre of water and add 1 ml of wetting agent (5).
(7) Buffer solution. To about 600 ml water, add in this sequence: 2 g of disodium ethylene diamine tetra acetate dihydrate, $Na_2EDTA \cdot 2H_2O$, 17.5 g of disodium hydrogen phosphate dihydrate, $Na_2HPO_4 \cdot 2H_2O$, and 24 g of sodium hydroxide, NaOH. After dissolution, make up to 1 litre and add 1 ml of wetting agent (5).
(8) Salicylate-nitroprusside solution. Dissolve 150 g of sodium salicylate, $C_7H_5O_3Na$, and 0.30 g of sodium nitroprusside dihydrate, Na_2-$[Fe(CN)_5NO] \cdot 2H_2O$, in about 800 ml water. Add 1 ml of wetting agent (5) and make up to 1 litre.
(9) Hypochlorite solution. A stock solution, containing approximately 1 M sodium hypochlorite in 0.1 M NaOH, should be purchased commercially. Dilute 20 ml of this stock solution to 200 ml. Prepare fresh daily.
(10) Mixed stock solution, N concentration 2500 mg/l, P concentration 500 mg/l, Na concentration 1250 mg/l and K concentration 4000 mg/l. Weigh out 11.793 g of ammonium sulphate, $(NH_4)_2SO_4$, 2.197 g of potassium dihydrogen phosphate, KH_2PO_4, 3.178 g of sodium chloride, NaCl, and 6.424 g of potassium chloride, KCl. Dissolve these in water in a 1000-ml volumetric flask and make up to the mark.

Remarks:
- Wash fluid 1 is used only once in a fortnight to rinse the N module because the N canal does not pollute so rapidly.

- In the P determination another wetting agent is used, because Brij 35 contains phosphate. Since the N and P canals cannot be rinsed separately, the N canal should be washed with the wash fluid that contains Aerosol 22.
- In the authors' laboratory, the mixed stock solution (10) is used because these four elements are determined in one run.
- The final solution of sodium hypochlorite (9) should contain 0.7% ± 0.1% of active chlorine. This will be true for solutions purchased from a supplier of chemicals, but if the hypochlorite was bought as bleach in a supermarket a check on its concentration is recommended.

Standard series

Pipette into 100-ml volumetric flasks, that contain already about 50 ml water, 4.5 ml of concentrated sulphuric acid (96%). Allow to cool. Then add 0–4.00–8.00–12.00 ml of the mixed stock solution (10) and dilute to volume.

Dilute this standard series 1 + 9 (v/v) in polycarbonate test tubes; this should be done together with the diluting of the digests. The diluted standard series has N concentrations of 0–10–20–30 mg/l.

Procedure

Dilute all digests 1 + 9 (v/v).

Make up a measuring series by placing standards and samples in the automatic sampler in the following sequence:

1–3 fourth term of the standard series (highest standard)
4 first term of the standard series (zero standard)
5 second term of the standard series
6 third term of the standard series
7 fourth term of the standard series (highest standard)
8 first term of the standard series (zero standard)
9 blank digestion
10–48 rest of the digestion series (i.e., 34 samples, two standard samples "low", two standard samples "high" and one more blank digestion)
49–50 fourth term of the standard series (highest standard).

Put the continuous-flow system into operation according to the scheme given herewith; rinse it during 5 minutes with wash fluid 2 (= reagent (4)) and then for 30 minutes with reagents; then put the sampler into operation.

Measure the absorbance at a wavelength of 660 nm. Plot a calibration curve and read the nitrogen concentrations.

After finishing the measurements (at the end of the day), flush the system for 15 min with wash fluid 2 (= reagent (4)).

Remarks:

- The highest standards at positions 1 and 2 serve to calibrate the recorder at a desired peak height. The standard at position 3, then, must give an equally high peak as the last two standards (49 and 50).

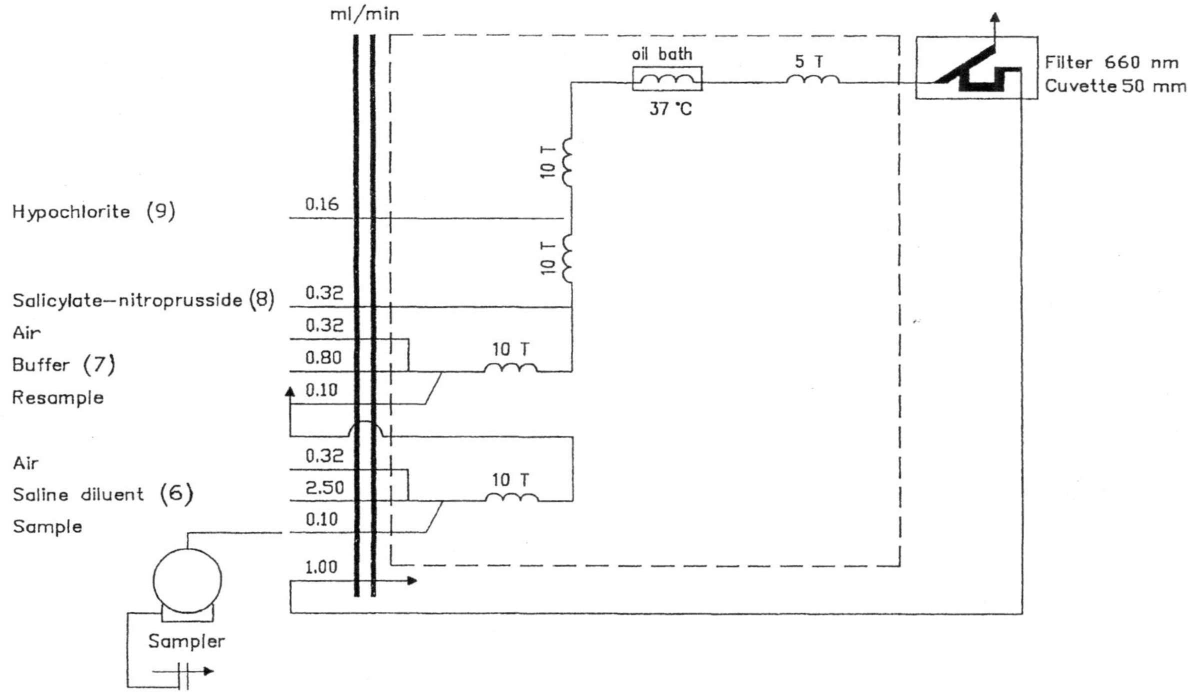

Flow diagram N-determination (1.5.3).

- The zero standard (position 8) may give too high a result because of influence of the highest standard measured just before. Therefore, this observation is rejected.

Calculation

The nitrogen content of the dried plant material, expressed in mmol/kg N, is calculated by

$$0.714 \times (a - b) \times \frac{V}{w}$$

in which
a = concentration of nitrogen in the diluted sample digest, in mg/l;
b = concentration of nitrogen in the diluted blank digest, in mg/l;
V = total volume of digest at the end of the digestion procedure, in ml;
w = weight of plant material sample, in g.

1.6.1 DETERMINATION OF **SODIUM** BY FLAME AES

Field of application

This determination is intended to be carried out on digest 1.0 (H_2SO_4-salicylic acid-H_2O_2-Se, in tubes).

The determination may be done in combination with K (1.2.1).

An automated procedure is described in section 1.6.2.

Principle

The sample is nebulized into an air-propane flame, where it is vaporized; sodium compounds are atomized and the sodium atoms thus formed emit radiation of which the intensity is measured at a wavelength of 589.0 nm.

Remarks:

- To prevent ionisation interferences, a cesium salt is added to act as an ionization buffer.
- Since the sodium atomic emission line finds itself in the wavelength region of a calcium oxide emission band, too high results will be found if calcium is present in the digest; however, since calcium in the sulphuric acid digests will form undissociated $CaSO_4$ in the flame, the interference of CaO is negligible.

Apparatus

- Flame emission spectrometer.

Reagents

(1) Cesium solution, Cs concentration 0.86 g/l. Dissolve 1.1 g of cesium chloride, CsCl, "pro analysi", in 1 litre of water.

(2) Stock solution, Na concentration 1250 mg/l. Dissolve 3.178 g of sodium chloride, NaCl, pretreated according to Appendix A-4, in some water in a 1000-ml volumetric flask and make up to the mark with water.

Remarks:

- The cesium chloride should be of the highest analytical grade ("pro analysi"), because a lower quality (e.g. "reinst" = "most pure") contains much more Na.
- Frequently, Na determinations go along with K determinations. In that case, a mixed stock solution may be prepared with 1250 mg/l Na and 4000 mg/l K (see 1.2.1 – reagent 2).

Standard series

Pipette 0–2.00–4.00–6.00–8.00–10.00–12.00 ml of the stock solution (2) into 100-ml volumetric flasks, to which already 50 ml water and 4.5 ml of concentrated sulphuric acid (96%) have been added in advance. Make up to the mark with water. Dilute this standard series 1 + 9 (v/v) with cesium solution (1); this should be done together with the diluting of the sample digests. This diluted standard series has Na concentrations of 0–2.5–5–7.5–10–12.5–15 mg/l.

Procedure

Dilute all digests 1 + 9 (v/v) with the cesium solution (1). Measure the emission intensity of the diluted sample digests, the diluted blank digests and the diluted standard series at a wavelength of 589.0 nm, using an air-propane flame. Plot a calibration curve and read the sodium concentrations.

Remarks:

- The calibration curve should be nearly linear.
- An air-acetylene flame can be used also, since the Cs concentration is high enough to counteract the greater tendency to ionisation.
- Instead of indicating a wavelength, a simple emission spectrometer ("flame photometer") may only be supplied with a so-called Na filter. Make sure that this is an interference filter, since glass filters are not selective enough.

Calculation

The sodium content of the dried plant material, expressed in mmol/kg Na, is calculated by

$$0.435 \times (a - b) \times \frac{V}{w}$$

in which

a = concentration of sodium in the diluted sample digest, in mg/l;
b = concentration of sodium in the diluted blank digest, in mg/l;
V = total volume of digest at the end of the digestion procedure, in ml;
w = weight of plant material sample, in g.

1.6.2 DETERMINATION OF **SODIUM** BY FLAME AES (AUTOMATED, BY FLOW ANALYZER)

Field of application

This determination is intended to be carried out on digest 1.0 (H_2SO_4-salicylic acid-H_2O_2-Se, in tubes).

The sodium content of the dried plant material should be higher than 25 mmol/kg.

The determination may be done in combination with the automated determination of K (1.2.2), N (1.5.3), and P (1.7.2) by flow analyzer.

The manual determination is described in section 1.6.1.

Principle

In a continuous-flow analysis (CFA) system, the sample is mixed with an internal standard (Li) and then vaporized in an air-propane flame. Sodium compounds are atomized; the sodium atoms thus formed emit radiation of which the intensity is measured at a wavelength of 589.0 nm.

Apparatus

- Continuous-flow analysis system (sampler, pump, K/Na unit, flame photometer). In the authors' laboratory, a Skalar Segmented Flow Analyzer is used, but any CFA system will do.

  Note: Depending on the flame photometer used, sodium can be determined really simultaneously with potassium in this automated procedure.
- Digital printer.

Reagents

(1) Wash fluid. Only demineralized water without any wetting agent should be used as wash fluid, in order to prevent contamination of the flame photometer by sodium.

(2) Internal standard. Dissolve 12.7 g of lithium chloride, LiCl, in 1000 ml water. Add 7.5 ml of this solution to about 800 ml water; add also 12.5 ml of glycerol, $C_3H_8O_3$, and make up to 1000 ml. Prepare just before use.

Remark:
- The internal standard may vary according to the flame photometer used. In the authors' laboratory, therefore, a "Radiometer solution" – supplied by the manufacturer of the flame photometer – is used as internal standard.

(3) Mixed stock solution, N concentration 2500 mg/l, P concentration 500 mg/l, Na concentration 1250 mg/l and K concentration 4000 mg/l. Weigh out 11.793 g of ammonium sulphate, $(NH_4)_2SO_4$, 2.1978 g of potassium

dihydrogen phosphate, KH_2PO_4, 3.178 g of sodium chloride, NaCl, and 6.424 g of potassium chloride, KCl. Dissolve these in bidest in a 1000-ml volumetric flask and make up to the mark with bidest.

Remarks:

- In the authors' laboratory, the mixed stock solution (3) is used because these four elements are determined in one run. Should the phosphate be left out, then more potassium chloride is needed for the potassium standard series.
- The mixed stock solution (3) and the standard series are made with bidest in favour of the P determination; for the potassium/sodium determination it is not necessary.

Standard series

Pipette into 100-ml volumetric flasks, that contain already about 50 ml water, 4.5 ml of concentrated sulphuric acid (96%). Allow to cool. Then add 0–4.00–8.00–12.00 ml of the mixed stock solution (3) and dilute to volume.

Dilute this standard series 1 + 9 (v/v) in polycarbonate test tubes with water; this should be done together with the diluting of the digests. The diluted standard series has Na concentrations of 0–5–10–15 mg/l.

Procedure

Dilute all digests 1 + 9 (v/v) with water in polycarbonate test tubes.

Make up a measuring series by placing standards and samples in the automatic sampler in the following sequence:

1–3 fourth term of the standard series (highest standard)
4 first term of the standard series (zero standard)
5 second term of the standard series
6 third term of the standard series
7 fourth term of the standard series (highest standard)
8 first term of the standard series (zero standard)
9 blank digestion
10–48 rest of the digestion series (i.e., 34 samples, two standard samples "low", two standard samples "high" and one more blank digestion)
49–50 fourth term of the standard series (highest standard).

Put the continuous-flow system into operation according to the scheme given herewith; rinse the system and the flame photometer during 5 minutes with demi water and next for 30 minutes with reagents; then put the sampler into operation.

Measure the Na emission intensity ratio with respect to Li. Print the Na-Li emission intensity ratios. Plot a calibration curve and calculate the Na concentrations.

After finishing the measurements (at the end of the day), flush the flame photometer for 15 min with water.

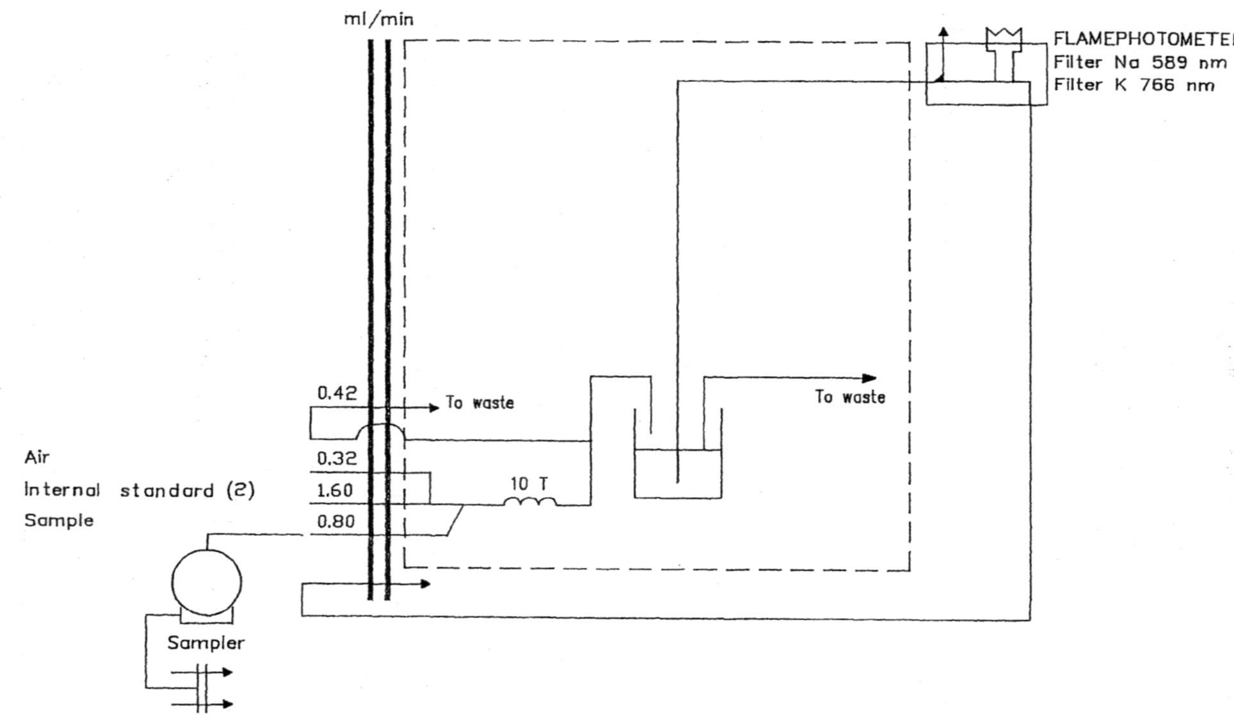

Flow Diagram Na-determination (1.6.2.).

Remarks:
- It is advantageous to use a printer instead of a recorder. The moment of printing should, however, be adjusted carefully so as to coincide with the top of the peak. Once adjusted, the tube length from the 10-turn coil to the flame photometer must remain constant.
- The standard at position 3 in the sampler should give the same value as those at positions 49 and 50.
- The zero standard (position 8) may give too high a result because of influence of the highest standard measured just before. Therefore, this observation is rejected.
- In the authors' laboratory, the sample stream is actually split up directly after the pump into three parallel streams: one for K/Na, one for N, one for P.

Calculation

The sodium content of the dried plant material, expressed in mmol/kg Na, is calculated by

$$0.435 \times (a - b) \times \frac{V}{w}$$

in which
a   concentration of sodium in the diluted sample digest, in mg/l;
b = concentration of sodium in the diluted blank digest, in mg/l;
V = total volume of digest at the end of the digestion procedure, in ml;
w = weight of plant material sample, in g.

1.7.1 DETERMINATION OF PHOSPHORUS BY COLORIMETRY

Field of application

This determination is intended to be carried out on digest 1.0 (H_2SO_4-salicylic acid-H_2O_2-Se, in tubes).

An automated procedure is described in section 1.7.2.

Principle

In an acidic medium, orthophosphates form a yellow-coloured complex with molybdate ions. After addition of ascorbic acid and Sb, a blue-coloured phosphomolybdenum complex is formed. Its absorbance is measured at a wavelength of 880 nm.

Remarks:
- As^{5+} and Si^{4+} form similarly coloured complexes under these conditions; As^{5+} interference, if any, may be suppressed by reduction of As^{5+} to As^{3+} (see remark with the procedure). Si^{4+} does not normally interfere with the digests in question, because these have been decanted.
- The blue colour varies with the redox conditions of the medium and with pH.
- Sb is added to complete the reduction of the yellow phosphomolybdenum complex to the blue phosphomolybdenum complex. Moreover, Sb enhances the intensity of the blue colour and allows a most sensitive measurement at 880 nm.

- With digest 1.0, the reducing agent forms elemental Se. An anti-coagulation agent must be added to keep the Se in colloidal state, so that no precipitate will interfere in the final measurement.

Apparatus

- Colorimeter, allowing measurements at 880 nm.

Reagents

(1) Sulphuric acid, $c(H_2SO_4)$ = 2.5 mol/l. Add carefully, while swirling, 140 ml of concentrated sulphuric acid (96%) to about 400 ml water. Allow to cool and make up to 1 litre.
(2) Sulphuric acid, $c(H_2SO_4)$ = 0.7 mol/l. Add carefully, while swirling, 40 ml of concentrated sulphuric acid (96%) to about 400 ml water. Allow to cool and make up to 1 litre.
(3) Molybdate solution. Dissolve 4 g of ammonium molybdate, $(NH_4)_6Mo_7O_{24} \cdot 4H_2O$ in 100 ml water. Store in a hard glass vessel.
(4) Ascorbic acid solution. Dissolve 1.76 g of ascorbic acid, $C_6H_8O_6$, in 100 ml water. Prepare fresh daily.
(5) Antimonyl tartrate solution. Dissolve 0.14 g of potassium antimonyl tartrate, $KSbC_4H_4O_7$, in 50 ml water.
(6) Anti-coagulation agent. Wetting agent Aerosol 22, to be purchased from Sigma Chemical Co., product number A 9753.
(7) Reduction mixture. Add 2 g of sodium metabisulphite, $Na_2S_2O_5$, and 0.2 g of sodium thiosulphate, $Na_2S_2O_3$, to about 200 ml water. Introduce a magnetic bar and stir until all crystals have dissolved. Add slowly, while still stirring, 10 ml of 2.5 M sulphuric acid (1); then make up to 250 ml. Prepare just before use.
(8) Mixed reagent. Add 15 ml of molybdate solution (3) to 50 ml of 2.5 M sulphuric acid (1) and mix. Next add 30 ml of ascorbic acid solution (4) and mix; then add 5 ml of antimonyl tartrate solution (5) and mix. Prepare fresh daily. See remark with procedure.
(9) Diluted mixed reagent. Mix 80 ml of the mixed reagent (8) with 300 ml water.
(10) Stock solution, P concentration 500 mg/l. Dissolve 2.197 g of potassium dihydrogen phosphate, KH_2PO_4, in about 400 ml water in a 1000-ml volumetric flask. Add carefully, while swirling, 40 ml of concentrated sulphuric acid (96%). Allow to cool and make up to the mark.

Remarks:
- The addition of sulphuric acid (1) to the reduction mixture (7) should be done slowly and while swirling, because otherwise the thiosulphate may decompose and form a precipitate of elemental S.
- The mixed stock solution of procedure 1.7.2 may also be used.

Standard series

Pipette 0–2.00–4.00–6.00–8.00–10.00 ml of the stock solution (10) into 100-ml volumetric flasks. Dilute to volume with 0.7 M sulphuric acid (2).

Dilute this standard series 1 + 9 (v/v) with water; this should be done together with the diluting of the digests. The diluted standard series has P concentrations of 0–1–2–3–4–5 mg/l.

Procedure

Dilute all digests 1 + 9 (v/v) with water.

Pipette 1.0 ml of the diluted sample digests, the diluted blank digests and the diluted standard series into test tubes. Add 3.8 ml of the diluted mixed reagent (9) and mix. Allow to stand for 1 hour. Measure the absorbance in a 1-cm cuvette at a wavelength of 880 nm. Plot a calibration curve and read the phosphorus concentrations.

Remarks:

- After having reached its maximum intensity, the colour is stable for at least 5 h.
- The absorbance can also be measured at 720 nm or 600 nm. The measurement is then somewhat less sensitive.
- Only part of the Mo^{6+} will be reduced. Hence, samples that have high phosphate concentrations cannot be diluted once that the blue colour has been formed.
- When As^{5+} is present, 3 ml of the reduction mixture (7) should be added to all solutions; after 15 minutes waiting, 0.8 ml of the *undiluted* mixed reagent (8) is added.

Calculation

The phosphorus content of the dried plant material, expressed in mmol/kg P, is calculated by

$$0.323 \times (a - b) \times \frac{V}{w}$$

in which

a = concentration of phosphorus in the diluted sample digest, in mg/l;
b = concentration of phosphorus in the diluted blank digest, in mg/l;
V = total volume of digest at the end of the digestion procedure, in ml;
w = weight of plant material sample, in g.

REFERENCES

J.Ch. van Schouwenburg and I. Walinga. The rapid determination of phosphorus in presence of arsenic, silicon and germanium. Anal. Chim. Acta *37* (1967) 271–274.

1.7.2 DETERMINATION OF **PHOSPHORUS** BY COLORIMETRY (AUTOMATED, BY FLOW ANALYZER)

Field of application

This determination is intended to be carried out on digest 1.0 (H_2SO_4-salicylic acid-H_2O_2-Se, in tubes).

The phosphorus content of the dried plant material should be higher than 20 mmol/kg.

The determination may be done in combination with the automated determination of K (1.2.2), N (1.5.3), and Na (1.6.2) by flow analyzer.

The manual determination is described in section 1.7.1.

Principle

Orthophosphate forms a yellow-coloured complex with molybdate ions in an acid medium. After addition of ascorbic acid and Sb, a blue-coloured phosphomolybdenum complex is formed, of which the absorbance is measured at a wavelength of 660 nm.

Remarks:

- As^{5+} and Si^{4+} form similarly coloured complexes under these conditions. As^{5+} interference is negligible at the low concentrations which are normally present in plant material. Si^{4+} does not normally interfere with the digests in question because these have been decanted after settling down of the silicates.
- Sb is added to complete the reduction of the yellow phosphomolybdenum complex to the blue phosphomolybdenum complex.

Apparatus

- Continuous-flow analysis system (sampler, pump, PO_4-unit, photometer, recorder). In the authors' laboratory, a Skalar Segmented Flow Analyzer is used, but any CFA system will do.

Reagents

For this determination, all reagents must be prepared with doubly distilled water *(bidest)* since demineralized water may contain some Si^{4+}.

(1) Sampler wash fluid. Bidest without any wetting agent.
(2) Wash fluid 1. Dissolve 4 g of sodium hydroxide, NaOH, in 1 litre of bidest.
(3) Wetting agent Aerosol 22, to be purchased from Sigma Chemical Co., product number A 9753.
(4) Wash fluid 2. Add 2 ml of wetting agent (3) to 1 litre of bidest.
(5) Saline diluent. Dissolve 20 g of sodium chloride, NaCl, in 1 litre of bidest and add 1 ml of wetting agent (3).
(6) Acid diluent. Dissolve 20 g of sodium chloride, NaCl, in about 400 ml of

bidest. Add carefully 50 ml of concentrated sulphuric acid (96%), allow to cool, add 1 ml of wetting agent (3) and make up to 1 litre with bidest.

(7) Ascorbic acid solution. Dissolve 10 g of ascorbic acid, $C_6H_8O_6$, in 1 litre of bidest. Prepare fresh weekly.

(8) Mixed reagent. Dissolve 30 g of ammonium molybdate tetrahydrate, $(NH_4)_6Mo_7O_{24} \cdot 4H_2O$, and 0.2 g of potassium antimonyl tartrate, $KSbC_4H_4O_7$, in about 400 ml of bidest. Add carefully 50 ml of concentrated sulphuric acid (96%) and allow to cool. Add 2 ml of wetting agent (3) and make up to 1 litre with bidest.

(9) Mixed stock solution, P concentration 500 mg/l, N concentration 2500 mg/l, Na concentration 1250 mg/l and K concentration 4000 mg/l. Weigh out 2.197 g of potassium dihydrogen phosphate, KH_2PO_4, 11.793 g of ammonium sulphate, $(NH_4)_2SO_4$, 3.178 g of sodium chloride, NaCl, and 6.424 g of potassium chloride, KCl. Dissolve these in bidest in a 1000-ml volumetric flask and make up to the mark with bidest.

Remarks:

- The sampler wash fluid (1) should match the digests, i.e., be 0.08 M sulphuric acid, but no difference in results has been found when using bidest.
- Since the phosphate salt brings already potassium in the system, less KCl needs to be weighed out than for a single stock solution.

Standard series

Pipette into 100-ml volumetric flasks, that contain already about 50 ml water, 4.5 ml of concentrated sulphuric acid (96%). Allow to cool. Then add 0–4.00–8.00–12.00 ml of the mixed stock solution (9) and dilute to volume.

Dilute this standard series 1 + 9 (v/v) in polycarbonate test tubes; this should be done together with the diluting of the digests. The diluted standard series has P concentrations of 0–2–4–6 mg/l.

Remark:

Whereas the reagents should be made in bidest, to prevent premature colouration, the standard series may be made in demineralized water. Moreover, all diluting may be done with demineralized water because any error from the use of this water is the same for both standards and samples.

Procedure

Dilute all digests 1 + 9 (v/v).

Make up a measuring series by placing standards and samples in the automatic sampler in the following sequence:

1–3	fourth term of the standard series (highest standard)
4	first term of the standard series (zero standard)
5	second term of the standard series
6	third term of the standard series

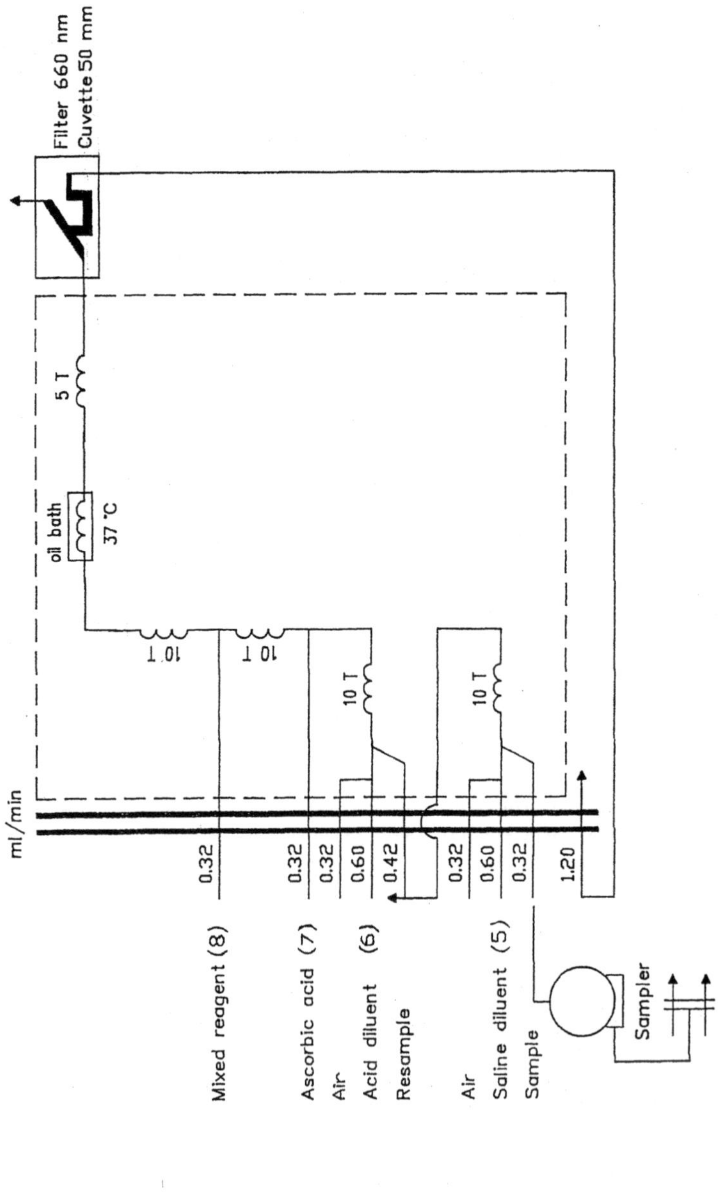

Flow diagram P-determination (1.7.2).

7 fourth term of the standard series (highest standard)
8 first term of the standard series (zero standard)
9 blank digestion
10–48 rest of the digestion series (i.e., 34 samples, two standard samples "low", two standard samples "high" and one more blank digestion)
49–50 fourth term of the standard series (highest standard).

Put the continuous-flow system into operation according to the scheme given herewith. Rinse it during 5 min with wash fluid 2 (= reagent (4)) and then for 30 min with reagents; then put the sampler into operation. Measure the absorbance at a wavelength of 660 nm. Plot a calibration curve and read the phosphorus concentrations.

After finishing the measurements (at the end of the day), flush the system during 15 min with wash fluid 1 (= reagent (2)), followed by 30 min with wash fluid 2 (= reagent (4)).

Remarks:

- The highest standards at position 1 and 2 serve to calibrate the recorder at a desired peak height. The standard at position 3, then, must give an equally high peak as the last two standards (49 and 50).
- The zero standard (position 8) may give too high a result because of influence of the highest standard measured just before. Therefore, this observation is rejected.
- The mixed reagent (8) must be connected to the system only when the rest is already running, otherwise a precipitate will be formed.
- The absorbance is measured at 660 nm because the system is provided with a filter of that wavelength. Measurements may, however, be taken at 720 nm or 880 nm as well.
- Since the determination of P goes together with N, this canal Is also flushed each day with wash fluid 1. This does not affect these measurements. The Na/K canal, however, is rinsed separately with demineralized water only.

Calculation

The phosphorus content of the dried plant material, expressed in mmol/kg P, is calculated by

$$0.329 \times (a - b) \times \frac{V}{w}$$

in which
a = concentration of phosphorus in the diluted sample digest, in mg/l;
b = concentration of phosphorus in the diluted blank digest, in mg/l;
V = total volume of digest at the end of the digestion procedure, in ml;
w = weight of plant material sample, in g.

1.8 DETERMINATION OF **ZINC** BY FLAME AAS

Field of application

This determination is intended to be carried out on digest 1.0 (H_2SO_4-salicylic acid-H_2O_2-Se, in tubes).

The determination may be done in combination with Mn (1.4).

Principle

The sample is nebulized into an air-acetylene flame, where it is vaporized; zinc compounds are atomized and the zinc atoms thus formed absorb radiation from a hollow-cathode lamp. The absorbance is measured at a wavelength of 213.9 nm.

Apparatus

- Atomic absorption spectrophotometer.

Reagents

(1) Sulphuric acid, $c(H_2SO_4)$ = 0.8 mol/l. Add carefully, while swirling, 45 ml of concentrated sulphuric acid (96%) to about 400 ml water. Allow to cool and make up to 1 litre.

(2) Stock solution, Zn concentration 400 mg/l. Dissolve 1.759 g of zinc sulphate heptahydrate, $ZnSO_4 \cdot 7H_2O$ (standardized according to Appendix A-24) in about 600 ml water in a 1000-ml volumetric flask. Add 45 ml of concentrated sulphuric acid (96%), allow to cool and make up to volume.

(3) Standard solution, Zn concentration 20 mg/l. Pipette 25.0 ml of the stock solution (2) into a 500-ml volumetric flask and make up to volume with 0.8 M sulphuric acid (1).

Remark:

- Frequently, Zn determinations go along with Mn determinations. In that case, a mixed standard solution may be prepared with 20 mg/l Zn and 50 mg/l Mn (see 1.4 – reagent 3).

Standard series

Pipette 0–1.00–2.00–4.00–6.00–8.00–10.00 ml of the standard solution (3) into 100-ml volumetric flasks and make up to volume with 0.8 M sulphuric acid (1). This standard series has Zn concentrations of 0–0.2–0.4–0.8–1.2–1.6–2.0 mg/l.

Procedure

Nebulize the sample digests, the blank digests and the standard series into a just blue (stoichiometric) air-acetylene flame and measure the absorbance at 213.9 nm. Plot a calibration curve and read the zinc concentrations.

Remarks:

- The calibration curve is slightly bent towards the x-axis, so that calculation by means of linear regression is not allowed.
- Although the chosen wavelength is short, no background correction has been found necessary when using 0.08 M sulphuric acid, 0.25 M sulphuric acid or 0.15 M perchloric acid. For other media, background correction may be required.
- No interferences were observed after spiking the digests with several salts, so as to reach final concentrations of 100 mg/l Mg, 30 mg/l Fe, 20 mg/l Al + 400 mg/l PO_4, or 30 mg/l Fe + 20 mg/l Al + 400 mg/l PO_4.
- Do not use rubber stoppers, since these will release zinc.

Calculation

The zinc content of the dried plant material, expressed in mg/kg Zn, is calculated by

$$(a - b) \times \frac{V}{w}$$

in which

a = concentration of zinc in the sample digest, in mg/l;
b = concentration of zinc in the blank digest, in mg/l;
V = total volume of digest at the end of the digestion procedure, in ml;
w = weight of plant material sample, in g.

2 Digestion in **flasks** with H_2SO_4-salicylic acid-H_2O_2 and determination of Ca, K, Mg, Mn, N, Na, P, Zn

2.0 DIGESTION IN **FLASKS** WITH H_2SO_4-SALICYLIC ACID-H_2O_2

Field of application

This digestion is intended for incidental, small series of plant samples. It can be applied for the determination of Ca, K, Mg, Mn, N-total, Na, P, and Zn in plant material.

Principle

With very strong sulphuric acid, the plant material is dehydrated and most of the organic matter oxidized at rather high temperature. The digestion is completed by hydrogen peroxide at elevated temperature.

Remarks:

- Salicylic acid is added to prevent loss of nitrate.

Apparatus

- Hot plate, able to reach 300 °C.
- Metal weighing funnel (see fig. 1).

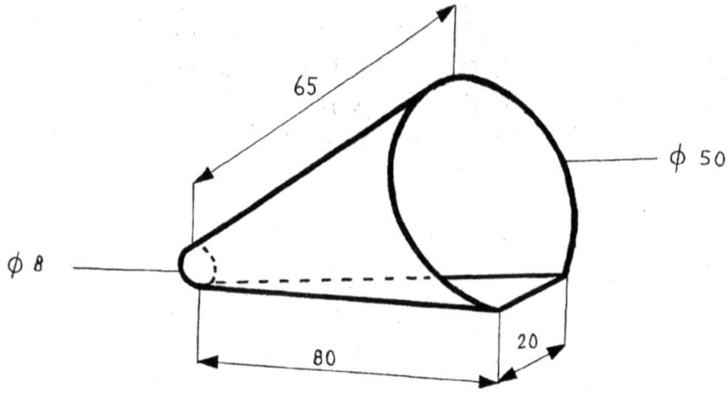

Fig. 1. Weighing funnel (measures in mm).

Remarks:
- Dried plant material may easily stick to glass when the relative humidity of the air is low. These home-made weighing funnels (stainless steel or aluminium) do not show this effect.

Reagents

(1) Sulphuric acid, 96% (w/w), $c(H_2SO_4) = 18$ mol/l (d = 1.84 g/ml).
(2) Hydrogen peroxide 30% (w/w).

Remark:
- Hydrogen peroxide must be of analytical quality; a lower grade may be stabilized with EDTA, phosphate, or other interfering compounds.

(3) Salicylic acid, powder.
(4) Digestion mixture. Put 18 ml water in a 250-ml erlenmeyer flask. While cooling, add in small portions 100 ml of sulphuric acid (1) **(CAUTION)**. Then dissolve 6 g of salicylic acid (3) with the aid of a magnetic stirrer.

Remark:
- Since the dried sample still contains about 10% (w/w) water, the sulphuric acid is somewhat diluted to prevent the temperature from raising too high during addition of the digestion mixture which otherwise would result in loss of nitrate (see remark in chapter 1.0).

Procedure

Weigh, to the nearest 0.001 g, approximately 0.3 g of the dried plant material sample in a metal weighing funnel and transfer the sample to a 50-ml volumetric flask. Take care that all the plant material comes below the neck of the flask. Add 3.3 ml of the digestion mixture (4) and swirl carefully until all the plant material is moistened. Allow to stand overnight. Prepare also two blank digestions.

Heat the flask on a hot plate at 180 °C for about 1 h. Remove the flask from the plate, let cool, and add 5 drops of hydrogen peroxide (2). Place the flask on the hot plate and increase the temperature to about 280 °C. Heat

5–10 min until the water has evaporated (appearance of white vapours). Remove the flask from the plate, let cool, add 5 drops of hydrogen peroxide (2), and heat again for 5–10 min to appearance of white vapours. Repeat this treatment until the digest has turned colourless.

Remove the flask from the plate and cool to room temperature. Add about 10 ml water and mix; swirl until most of the precipitate has dissolved. Make up to the mark with water, mix well and filter over coarse filter paper.

Remarks:

- After moistening the sample with the digestion mixture (4), at least 2 h are needed to form the nitro-salicylic acid compounds. A longer period (overnight), however, will prevent foaming later on.
- When heating at 180 °C, the sample will turn black and foam may be formed. If this reaches the neck of the flask, 1 or 2 drops of hydrogen peroxide (2) should be added.
- A precipitate may be formed when cooling after completing the digestion. The addition of 10 ml water then produces enough heat to dissolve the precipitate rapidly.
- The digest is filtered to remove any SiO_2 that will otherwise dissolve gradually and then interfere in the determinations.
- Do not use the volumetric flasks for other purposes.
- Near the end of the digestion, the liquid becomes yellow due to the last portions of organic matter, or to Fe. If the yellow colour persists after extra addition(s) of H_2O_2, then the presence of Fe should be assumed and the digestion considered as completed.
- The sulphate concentration in the blank digest is 0.8 mol/l.

2.1.1 DETERMINATION OF **CALCIUM** BY FLAME AES

Field of application

This determination is intended to be carried out on digest 2.0 (H_2SO_4-salicylic acid-H_2O_2, in flasks).

With calcium concentrations higher than 1 mol/kg dry material, the solubility product of $CaSO_4$ can be surpassed in the digest; a smaller sample has to be weighed out then.

Principle

The sample is vaporized in an air-acetylene flame; calcium compounds are atomized and the CaOH molecules thus formed emit radiation of which the intensity is measured at a wavelength of 622 nm.

Remarks:

- Although Ca measurements are also possible at 423 nm and 554 nm, the spectral interferences at 622 nm are smaller for elements present in plant material, whereas larger interferences at this wavelength only come from elements that hardly or not occur in plant material.
- When using the cooler air-propane flame, the emission signal is 30–40% lower, and there will be more interference.

Apparatus

- Flame emission spectrometer.

Reagents

(1) Hydrochloric acid, c(HCl) = 4 mol/l. Add 340 ml of concentrated hydrochloric acid (36%) to about 400 ml water and make up to 1 litre.
(2) Stock solution, Ca concentration 5000 mg/l. Weigh out 12.486 g of calcium carbonate, $CaCO_3$, pretreated according to Appendix A-4. Transfer it to a 1000-ml volumetric flask with the help of about 150 ml water. Add 63 ml of 4 M hydrochloric acid (1) and boil to expel CO_2. If calcium carbonate particles remain visible, add 5 ml of 4 M hydrochloric acid (1) extra. Allow to cool and make up to volume with water.

Standard series

Pipette 0–2.00–4.00–6.00–8.00–10.00 ml of the stock solution (2) into 100-ml volumetric flasks, to which already 50 ml water and 4.5 ml of concentrated sulphuric acid (96%) per flask have been added in advance. Make up to the mark with water.

Dilute this standard series 1 + 9 (v/v) with water; this should be done together with the diluting of the sample digests. This diluted standard series has Ca concentrations of 0–10–20–30–40–50 mg/l.

Procedure

Dilute all digests 1 + 9 (v/v) with water.

Measure the emission intensity of the diluted digests, the diluted blank digests and the diluted standard series at a wavelength of 622 nm, using an air-acetylene flame. Plot a calibration curve and read the calcium concentrations.

Remarks:

- The calibration curve should be linear, but shows a slight bending towards the x-axis. This is due to ionisation, which could be counteracted by addition of a cesium salt. The bending is, however, so small that Cs addition is not necessary.
- In determination 1.1.2 a releasing agent is added to prevent calcium atoms from being caught into poorly dissociating compounds, in particular oxides. Such an addition is not necessary here, because calcium is measured at an oxide band.
- When an air-propane flame is used, there may be interference from Fe and Al, which is counteracted by addition of magnesium chloride. This causes, however, such a higher salt level that the nebulizer may readily be clogged; therefore, water should be nebulized very frequently.
- Instead of indicating a wavelength, a simple emission spectrometer ("flame photometer") may only be supplied with a so-called Ca filter. Make sure that this is an interference filter, since glass filters are not selective enough.

Calculation

The calcium content of the dried plant material, expressed in mmol/kg Ca, is calculated by

$$0.2495 \times (a - b) \times \frac{V}{w}$$

in which
a = concentration of calcium in the diluted sample digest, in mg/l;
b = concentration of calcium in the diluted blank digest, in mg/l;
V = total volume of digest at the end of the digestion procedure, in ml;
w = weight of plant material sample, in g.

2.1.2 DETERMINATION OF **CALCIUM** BY FLAME AAS

Field of application

This determination is intended to be carried out on digest 2.0 (H_2SO_4-salicylic acid-H_2O_2, in flasks).

With calcium concentrations higher than 1 mol/kg dry material, the solubility product of $CaSO_4$ can be surpassed in the digest; a smaller sample has to be weighed out then.

Principle

The sample is vaporized in an air-acetylene flame; calcium compounds are atomized and the calcium atoms thus formed absorb radiation from a hollow-cathode lamp. The absorbance is measured at a wavelength of 422.7 nm.

Remark:
- Since Ca atoms are easily captured in the flame into poorly dissociating compounds (like oxides), a releasing agent like lanthanum must be added.

Apparatus

- Atomic absorption spectrophotometer.

Reagents

(1) Hydrochloric acid, c(HCl) = 4 mol/l. Add 340 ml of concentrated hydrochloric acid (36%) to about 400 ml water and make up to 1 litre.
(2) Lanthanum solution, La concentration 1 g/l. Dissolve 3.12 g of lanthanum nitrate hexahydrate, $La(NO_3)_3 \cdot 6H_2O$, in some water and make up to 1 litre with water.
(3) Stock solution, Ca concentration 1000 mg/l. Weigh out 2.497 g of $CaCO_3$

(pretreated according to App. A-4) and transfer it with 100–150 ml water to a 1-litre volumetric flask. Add 13 ml of 4 M hydrochloric acid (1) and boil to expel CO_2. If calcium carbonate particles remain visible, add 1 ml of 4 M hydrochloric acid (1) extra. Allow to cool and make up to volume with water.

Standard series

Pipette 0–2.00–4.00–6.00–8.00–10.00 ml of the stock solution (3) into 100-ml volumetric flasks, to which already 50 ml water and 4.5 ml of concentrated sulphuric acid (96%) per flask have been added in advance. Make up to the mark with water. This standard series has Ca concentrations of 0–20–40–60–80–100 mg/l.

Procedure

Pipette 0.25 ml of the sample digests, the blank digests and the standard series into test tubes. Add 4.75 ml of lanthanum solution (2). Nebulize each measuring solution into a yellow (reducing) air-acetylene flame and measure the absorbance at a wavelength of 422.7 nm. Plot a calibration curve and read the calcium concentrations.

Remark:

- The calibration curve should be linear by virtue of the releasing action of lanthanum.

Calculation

The calcium content of the dried plant material, expressed in mmol/kg Ca, is calculated by

$$0.0250 \times (a - b) \times \frac{V}{w}$$

in which
a = concentration of calcium in the sample digest, in mg/l;
b = concentration of calcium in the blank digest, in mg/l;
V = total volume of digest at the end of the digestion procedure, in ml;
w = weight of plant material sample, in g.

2.1.3 DETERMINATION OF **CALCIUM** BY COMPLEXOMETRIC TITRATION

Field of application

This determination is intended to be carried out on digest 2.0 (H_2SO_4-salicylic acid-H_2O_2, in flasks).

This titration is only recommended in case no flame AES or flame AAS is available.

Principle

Calcium ions are titrated with EDTA using HHSNN as a combined metal ion/acid-base indicator.

Remarks:

- Interfering ions are masked by cyanide and by triethanolamine.
- With magnesium ions added, a much clearer end point is obtained; the magnesium hydroxide formed is kept in colloidal state by a colloid protector.
- HHSNN is a very sensitive indicator, which – in the presence of Ca and Mg – is coloured red-violet from pH 4–8.5, blue from pH 9–9.5, red-violet from pH 10.5–11.5, blue-violet at pH 11.7 and red at pH 12.1 and higher.

Apparatus

- Magnetic stirrer.
- Plunger-type burette, or 25-ml burette.

Reagents

(1) Sulphuric acid, $c(H_2SO_4)$ = 0.7 mol/l. Add carefully, while swirling, 4 ml of concentrated sulphuric acid (96%) to about 40 ml water. Allow to cool and make up to 100 ml.
(2) Hydrochloric acid, c(HCl) = 1.0 mol/l. Add 8.3 ml of concentrated hydrochloric acid (36%) to about 40 ml water and make up to 100 ml.
(3) Potassium hydroxide solution, c(KOH) = 2 mol/l. Dissolve 56.1 g of potassium hydroxide, KOH, in about 400 ml of water. Allow to cool and make up to 500 ml.
(4) Magnesium additive. Dissolve 0.384 g of magnesium sulphate heptahydrate, $MgSO_4 \cdot 7H_2O$, in about 450 ml of water in a beaker. Boil the solution and add, while stirring, 12.5 g of sodium carboxymethylcellulose (for example, Nymcel ZLF 20; see remark). Remove the beaker from the source of heat and continue stirring until the larger particles have dissolved and only a slightly turbid solution is remaining. Allow to cool further and make up to 500 ml. Filter on a Büchner funnel over medium filter paper. When the filtration speed ceases too much, a new filter paper should be installed.
(5) Masking solution. Transfer 31 ml of triethanolamine, $C_6H_{15}NO_3$, to a 250-ml volumetric flask. Because of its high viscosity, this should be done by means of 3 portions of water of about 40 ml each. Then add 4.0 g of potassium cyanide, KCN, (CAUTION!) to the solution, dissolve it and make up to volume. Mix thoroughly.
(6) Indicator. Mix 20 g of sodium chloride, NaCl, with 100 mg of 2-hydroxy-

1-(2-hydroxy-4-sulpho-1-naphtylazo)-3-naphtoic acid, $C_{21}H_{14}N_2O_7S$, (abbreviated to HHSNN).

(7) EDTA solution, 0.0025 mol/l. Dissolve 0.93 g of ethylene diamine tetra acetic acid disodium salt, Na_2EDTA, in 1 litre of water. Standardize according to procedure (a) below.

(8) Standard solution, Ca concentration 0.00250 mol/l. Transfer 0.2502 g of calcium carbonate, $CaCO_3$, with the help of about 200 ml water into a 1000-ml volumetric flask. Add 6 ml of 1 M HCl (2) and boil to expel carbon dioxide. Allow to cool. Add 40 ml of concentrated sulphuric acid (96%), allow to cool again and make up to volume.

Remarks:

- The indicator HHSNN is also known as Cal-red and as Calcon-carboxylic acid.
- Application of polyvinyl alcohol also prevents the formation of precipitates, but in this case no colour change is shown anymore.

Standard series

With this titration no standard series is required, but the EDTA solution (7) should be standardized against the calcium standard solution (8) in exactly the same way as the sample digests are titrated.

Procedure

(a) Standardization of EDTA

Pipette A ml ($5.00 \leq A \leq 20.00$, depending on the capacity of the burette) of the calcium standard solution (8) and titrate it following procedure (b). Perform this titration in triplicate.

Perform also a "blank" (zero standard) by titrating A ml of 0.7 M sulphuric acid (1), also in triplicate.

Calculate the titer of the EDTA solution (7) by

$$t = 0.00250 \times \frac{A}{(p - q)}$$

in which

t = titer (= concentration) of the EDTA solution, in mol/l;

A = volume of calcium standard solution (8) pipetted, in ml;

p = average volume of EDTA solution used for titration of the standard solution, in ml;

q = average volume of EDTA solution used for titration of the zero standard solution, in ml;

(b) Titration of digests

Pipette D ml (D varying between 1 and 20, depending on the expected amount of calcium, which should be at most 1.5 mg) of the sample digest into a 100-ml beaker. Put the beaker on a magnetic stirrer and insert a magnetic stirring bar. Start stirring and continue it until the end of the titration.

Add 4.0 ml of magnesium additive (4) and a very small knife-point (10 mg) of indicator (6). *(Note: in the still acidic medium the indicator hardly dissolves. Therefore stop the addition of indicator when a faint red colour is observed.)* Add dropwise potassium hydroxide solution (3) to neutralize the acid and to dissolve the indicator. Proceed until the colour changes from faint red via practically colourless or slightly yellow to blue (pH 9–10). Then add 1 ml of masking solution (5), followed by 4 ml of potassium hydroxide solution (3); the solution should now show a red colour. Add water until the total volume of liquid in the beaker is about 40 ml.

Titrate with standardized 0.0025 M EDTA solution (7) until the indicator does not show a red hue anymore. The colour should now be blue, but because of the masking discolourations might show.

Titrate a blank digest in an identical manner.

Record the burette readings to the nearest 0.01 ml.

Remarks:

- With digest 2.0 no interferences were observed from 0.2 mg Al, 0.01 mg Co, 0.01 mg Cu, 0.5 mg Fe, 0.02 mg Ni, 5 mg PO_4, or 0.2 mg Zn (amounts present in D ml of digest). A slight lowering of the results was observed with 1.5 mg Mg or 1.5 mg Mn present.
- If Nymcel ZLF 20 is not available, Nymcel ZYF 30 can be used (to be purchased from Nyma, P.O. Box 31, 6500 AA Nijmegen, The Netherlands).

Calculation

The calcium content of the dried plant material, expressed in mmol/kg Ca, is calculated by

$$1000 \times (a - b) \times \frac{t}{D} \times \frac{V}{w}$$

in which

a = volume of EDTA solution used for titrating the sample digest, in ml;
b = volume of EDTA solution used for titrating the blank digest, in ml;
t = titer of the EDTA solution, in mol/l;
D = volume of digest taken for titration, in ml;
V = total volume of digest at the end of the digestion procedure, in ml;
w = weight of plant material sample, in g.

2.2 DETERMINATION OF **POTASSIUM** BY FLAME AES

Field of application

This determination is intended to be carried out on digest 2.0 (H_2SO_4-salicylic acid-H_2O_2, in flasks).

The determination may be done in combination with Na (2.6).

Principle

The sample is nebulized into an air-propane flame, where it is vaporized; potassium compounds are atomized and the potassium atoms thus formed emit radiation of which the intensity is measured at a wavelength of 766.5 nm.

Remark:

- To prevent ionisation interferences, a cesium salt is added, to act as an ionisation buffer.

Apparatus

- Flame emission spectrometer.

Reagents

(1) Cesium solution, Cs concentration 0.87 g/l. Dissolve 1.1 g of cesium chloride, CsCl, "pro analysi", in 1 litre of water.
(2) Stock solution, K concentration 4000 mg/l. Dissolve 7.627 g of potassium chloride, KCl, in some water in a 1000-ml volumetric flask and make up to the mark with water.

Remarks:

- The cesium chloride should be of the highest analytical grade ("pro analysi"), because a lower quality (e.g. "reinst" = "most pure") contains much more Na.
- Frequently, K determinations go along with Na determinations. In that case, a mixed stock solution may be prepared with 4000 mg/l K and 1250 mg/l Na (see 2.6 – reagent 2).

Standard series

Pipette 0–2.00–4.00–6.00–8.00–10.00–12.00 ml of the stock solution (2) into 100-ml volumetric flasks, to which already 50 ml water and 4.5 ml of concentrated sulphuric acid (96%) per flask have been added in advance. Make up to the mark with water.

Dilute this standard series 1 + 9 (v/v) with the cesium solution (1); this should be done together with the diluting of the sample digests. This diluted standard series has K concentrations of 0–8–16–24–32–40–48 mg/l.

Procedure

Dilute all digests 1 + 9 (v/v) with the cesium solution (1).

Measure the emission intensity of the diluted sample digests, the diluted blank digests and the diluted standard series at a wavelength of 766.5 nm, using an air-propane flame. Plot a calibration curve and read the potassium concentrations.

Remarks:

- The calibration curve should be nearly linear.

- An air-acetylene flame can be used also, since the Cs concentration is high enough to counteract the greater tendency to ionisation.
- Instead of indicating a wavelength, a simple emission spectrometer ("flame photometer") may only be supplied with a so-called K filter. Make sure that this is an interference filter, since glass filters are not selective enough.
- When using a filter in the spectrometer, the emission from K at the wavelength of 769.9 nm is also measured. This enhances the sensitivity of the K measurement.

Calculation

The potassium content of the dried plant material, expressed in mmol/kg K, is calculated by

$$0.256 \times (a - b) \times \frac{V}{w}$$

in which
a = concentration of potassium in the diluted sample digest, in mg/l;
b = concentration of potassium in the diluted blank digest, in mg/l;
V = total volume of digest at the end of the digestion procedure, in ml;
w = weight of plant material sample, in g.

2.3.1 DETERMINATION OF **MAGNESIUM** BY FLAME AAS

Field of application

This determination is intended to be carried out on digest 2.0 (H_2SO_4-salicylic acid-H_2O_2, in flasks).

This procedure is intended for incidental use; when Mg is to be determined more frequently and in large series, digestion 1.0 followed by determination 1.3 is recommended.

Principle

The sample is nebulized into an air-acetylene flame, where it is vaporized; magnesium compounds are atomized and the magnesium atoms thus formed absorb radiation from a hollow-cathode lamp. The absorbance is measured at a wavelength of 285.2 nm.

Remark:

- Since Mg atoms are easily captured in the flame into poorly dissociating compounds, in particular with phosphate and aluminium, a releasing agent like lanthanum must be added.

Apparatus

- Atomic absorption spectrophotometer, in which the burner can be turned with respect to the light beam.

Reagents

(1) Sulphuric acid, $c(H_2SO_4)$ = 0.08 mol/l. Add 4.5 ml of concentrated sulphuric acid (96%) to about 400 ml water and make up to 1 litre.
(2) Lanthanum solution, La concentration 10 g/l. Dissolve 7.78 g of lanthanum nitrate hexahydrate, $La(NO_3)_3 \cdot 6H_2O$, in some water and make up to 250 ml.
(3) Standard solution, Mg concentration 40 mg/l. Dissolve 0.4056 g of magnesium sulphate heptahydrate, $MgSO_4 \cdot 7H_2O$, standardized according to App. A-20, in some water in a 1000-ml volumetric flask. Dilute to about 0.5 litre, add 4.5 ml of concentrated sulphuric acid (96%) and make up to volume.

Remark:
- Since this determination is expected to be done only now and then, the standard solution should have a long shelf life. Therefore acid is added.

Standard series

Pipette 0–1.00–2.00–4.00–6.00–8.00–10.00 ml of the standard solution (3) into 100-ml volumetric flasks. Add 10 ml of lanthanum solution (2) and make up to volume with 0.08 M sulphuric acid (1). This standard series has Mg concentrations of 0–0.4–0.8–1.6–2.4–3.2–4.0 mg/l.

Procedure

Pipette 1.00 ml of the sample digests and the blank digests into test tubes. Add 8.00 ml of water and mix well. Then add 1.00 ml of lanthanum solution (2) and mix again. Nebulize the highest standard into a just blue (stoichiometric) air-acetylene flame and measure its absorbance at 285.2 nm. If necessary, turn the burner until the absorbance does not exceed the value of 0.6 anymore, and fix the burner at this position. Then nebulize the diluted digests and the standard series into the flame and measure the absorbance. Plot a calibration curve and read the magnesium concentrations.

Remarks:
- The calibration curve is slightly bent towards the x-axis.
- No interferences were observed after spiking the digests with 200 mg/l Al, 600 mg/l PO_4 or 200 mg/l SO_4.

Calculation

The magnesium content of the dried plant material, expressed in mmol/kg Mg, is calculated by

$$0.411 \times (a - b) \times \frac{V}{W}$$

in which
a = concentration of magnesium in the diluted sample digest, in mg/l;
b = concentration of magnesium in the diluted blank digest, in mg/l;
V = total volume of digest at the end of the digestion procedure, in ml;
w = weight of plant material sample, in g.

2.3.2 DETERMINATION OF **MAGNESIUM** BY COLORIMETRY

Field of application

This determination is intended to be carried out on digest 2.0 (H_2SO_4-salicylic acid-H_2O_2), in flasks.

This colorimetric determination is laborious and requires extra attention; therefore, it is recommended only in case no flame AAS apparatus is available.

Principle

Under carefully controlled circumstances, the sample solution is made alkaline so that magnesium hydroxide is formed. This is adsorbed by titan yellow, thus forming a red-coloured so-called lake of which the absorbance is measured at a wavelength of 550 nm.

Remarks:

The magnesium hydroxide formed should be kept in a colloidal state. Therefore, a mixture of polyvinylalcohol and glycerol is added to serve as a colloid protector.

- Phosphate is used to mask several metal ions, like Fe and Al.
- Calcium ions may render the determination irreproducible. This is prevented by addition of the chelating agent EGTA to the solution. Besides, the EGTA prevents the formation of any calcium phosphate.
- Different batches of titan yellow may show large variations in their sensitivity of reaction with magnesium. Therefore, each batch must be standardized (see Procedure).
- In this determination procedure, use is made of a so-called compensation solution (see under Reagents). This solution is added to compensate for unintended Mg-complexation (see remark) and to eliminate the influence of variations in interfering metal ion concentrations (see remark).

Apparatus

- Colorimeter with interference filter.

Reagents

(1) Sulphuric acid, $c(H_2SO_4)$ = 0.36 mol/l. Add carefully, while swirling, 20 ml of concentrated sulphuric acid (96%) to about 400 ml water. Allow to cool and make up to 1 litre.

(2) Hydrochloric acid, c(HCl) = 1.2 mol/l. Add 100 ml of concentrated hydrochloric acid (36%) to about 400 ml water and make up to 1 litre.

(3) Potassium hydroxide solution, c(KOH) = 6 mol/l. Dissolve 168 g of potassium hydroxide, KOH, carefully in about 400 ml water. Stir with a rod until dissolved. Allow to cool and make up to 500 ml.
(4) Phosphate solution. Dissolve 1.75 g of potassium dihydrogen phosphate, KH_2PO_4, in 100 ml water.
(5) Manganese solution. Dissolve 0.101 g of manganese sulphate tetrahydrate, $MnSO_4 \cdot 4H_2O$, in some water and make up to 100 ml.
(6) Aluminium solution. Dissolve 0.111 g of aluminium chloride hexahydrate, $AlCl_3 \cdot 6H_2O$, in some water and make up to 100 ml.
(7) Magnesium additive. Dissolve 0.101 g of magnesium sulphate heptahydrate, $MgSO_4 \cdot 7H_2O$, in some water and make up to 100 ml.
Note: This is not a standard solution.
(8) EGTA solution. Dissolve 12.5 g of ethylene glycol bis(2-aminoethyl-ether)-N,N,N',N'-tetra-acetic acid, EGTA, in 50 ml water and the minimum amount (about 14 ml) of potassium hydroxide solution (3). Neutralize to pH 7 with sulphuric acid (1) with the help of indicator paper or a pH meter.
(9) Compensation solution. Transfer the whole EGTA solution (8) to a 500-ml volumetric flask and add 25 ml of triethanolamine. Add subsequently all of the manganese solution (5), all of the aluminium solution (6), and 12.5 ml of the magnesium additive (7); then make up to volume. Prepare every month.
(10) Polyvinylalcohol solution. Weigh out 1 gram of polyvinylalcohol, average molecular weight approximately 14 000, in a beaker. Dissolve the polyvinylalcohol by boiling (carefully, because of formation of foam). Cool and dilute to 100 ml.
(11) Titan yellow solution. Dissolve 100 mg of titan yellow (thiazol yellow, C.I. 19540) in 100 ml water. Prepare fresh daily. See remark.
(12) Titan yellow reagent. Mix 10 ml of titan yellow solution (11) with 5 ml of polyvinylalcohol solution (10) and 25 ml of phosphate solution (4). Make up with water to 100 ml. Prepare fresh daily.
(13) Mixed reagent. Mix equal volumes of titan yellow reagent (12) and compensation solution (9).
(14) Stock solution, Mg concentration 100 mg/l. Dissolve 1.014 g of magnesium sulphate heptahydrate, $MgSO_4 \cdot 7H_2O$, standardized according to Appendix A-20, in about 200 ml water in a 1000-ml volumetric flask. Add carefully, while swirling, 20 ml of concentrated sulphuric acid (96%). Allow to cool and make up to volume.

Remarks:

- EGTA will bind $Ca2^{+}$ ions while leaving Mg^{2+} ions free. However, EGTA is often contaminated with other complexing compounds which may catch part of the Mg. In order to compensate for this effect, a small extra amount of Mg is added. Moreover, in this way, the solubility product of $Mg(OH)_2$ is also reached for samples and standard solutions that are very low in Mg. (This is necessary because Mg will react only in form of $Mg(OH)_2$ with titan yellow).

- Triethanolamine is added in order to complex Al- and Mn-ions. A rest influence of these two ions is equalized by the compensation solution (9).
- Potassium hydroxide is preferred over sodium hydroxide, because the former is purer. In the sulphuric acid digests (2.0), then, a precipitate of K_2SO_4 might be formed. Therefore, these digests are diluted.
- From the two types of titan yellow sold, only the type mentioned as reagent for the Mg determination should be used, because the composition of titan yellow produced by various manufacturers – and even of different batches from the same manufacturer – may show great differences in behaviour. A standard series should be made to see whether the concentration of titan yellow mentioned here (100 mg per 100 ml) will result in a good curve (see Procedure (a)). Making use of a recent titan yellow batch (from Merck), the optimum titan yellow concentration could be decreased to 25 mg per 100 ml.

Standard series

Pipette 0–1.00–2.00–3.00–4.00–5.00–6.00–8.00–10.00 ml of the stock solution (14) into 100-ml volumetric flasks. Make up to 100 ml with 0.36 M sulphuric acid (1). This standard series has Mg concentrations of 0–1–2–3–4–5–6–8 –10 mg/l.

Procedure

(a) Optimization of the standard series

When performing the method *for the first time* or running it with a *new batch* of EGTA or titan yellow (compare the lot numbers), the following procedure is recommended:

Prepare a standard series, and measure the absorbance according to procedure (b). Plot the calibration curve and inspect it carefully.

A good curve should *not* show a horizontal part for the lowest concentrations of Mg; the zero standard should read about 0.25–0.28 absorbance units and the curve should only start levelling off for the two highest standards. Such a curve is shown in Figure 2.

Three possible deviations from the best curve may be observed:

(i) The curve runs horizontally at the lowest Mg concentrations. In this case, the compensation solution (9) should be given an extra amount of magnesium; this is best done by addition of a small volume of magnesium additive (7). This volume can be estimated from the length of the horizontal part of the curve.

(ii) The curve has both a slight slope and a large intercept, and strongly levels off for the highest standards. In this case, there is too much magnesium in the compensation solution (9). The concentration of the magnesium additive should be lowered, and a new compensation solution prepared.

(iii) The curve has the right shape, but all standards show too high an absorbance. This may be the case when a new batch of titan yellow has a higher content of active substance. In this case, the concentration of

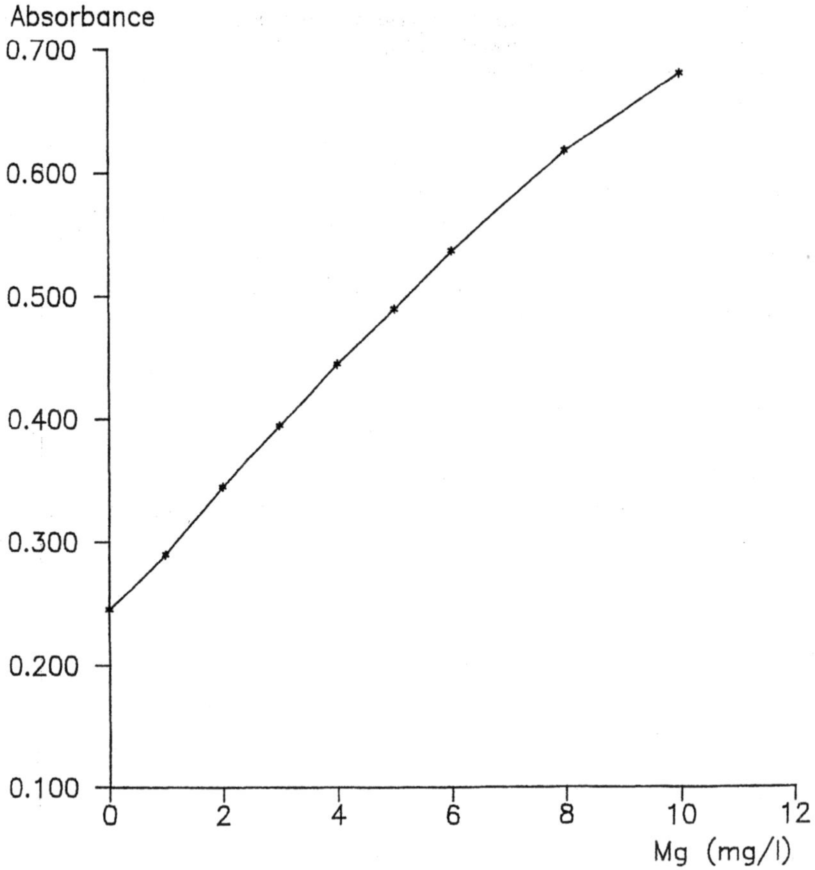

Fig. 2. Idealized calibration curve for colorimetric Mg determination.

the titan yellow solution (11) should be lowered, and a new calibration curve prepared.

(b) Measurement

Dilute in advance all digests 1 + 1 with water.

Pipette 2.00 ml of the standard series and of the diluted digests into test tubes. For duplicate measurements, pipette 1.00 ml of the diluted digests into test tubes and add 1.00 ml of 0.36 M sulphuric acid (1).

Add 2.00 ml of the mixed reagent (13); mix well. Then add 1.00 ml of potassium hydroxide solution (3) and mix thoroughly. Wait for 1 hour, while keeping the solutions in the dark. Mix again just before measurement, and measure the absorbance at a wavelength of 550 nm. Plot a calibration curve and read the magnesium concentrations.

Remarks:

- After the 1-h period of development, the colour is constant for at least 3 hours.

- For digest 2.0, no interferences were found after spiking the digests with metal ions to reach the following concentrations: 100 mg/l Al, 2500 mg/l Ca, 150 mg/l Co, 150 mg/l Cu, 150 mg/l Fe^{3+}, 150 mg/l Mn, 150 mg/l Ni, 75 mg/l Zn.

Calculation

The magnesium content of the dried plant material, expressed in mmol/kg Mg, is calculated by

$$0.0823 \times (a - b) \times f \times \frac{V}{w}$$

in which
a = concentration of Mg in the diluted sample digest, in mg/l;
b = concentration of Mg in the diluted blank digest, in mg/l;
f = dilution factor for the duplicate measurements; (f = 2 for 1 ml pipetted)
V = total volume of digest at the end of the digestion procedure, in ml;
w = weight of plant material sample, in g.

REFERENCES

E.G. Bradfield. A study of some factors which affect the adsorption of titan yellow on magnesium hydroxide. Anal.Chim.Acta *27* (1962) 262–271.
J.Ch. van Schouwenburg. The determination of magnesium with titan yellow; masking interferences with phosphate. Neth. J. agric. Sci. *13* (1965) 53–56.

2.4 DETERMINATION OF **MANGANESE** BY FLAME AAS

Field of application

This determination is intended to be carried out on digest 2.0 (H_2SO_4-salicylic acid-H_2O_2, in flasks).

The manganese content of the dried plant material should be higher than 50 mg/kg.

The determination may be carried out in combination with Zn (2.8).

Principle

The sample is nebulized into an air-acetylene flame, where it is vaporized; manganese compounds are atomized and the manganese atoms thus formed absorb radiation from a hollow-cathode lamp. Lanthanum is added to prevent condensed phase interferences. The absorbance is measured at a wavelength of 279.5 nm.

Apparatus

- Atomic absorption spectrophotometer.

Reagents

(1) Sulphuric acid, $c(H_2SO_4) = 0.8$ mol/l. Add carefully, while swirling, 45 ml of concentrated sulphuric acid (96%) to about 400 ml water. Allow to cool and make up to 1 litre.

(2) Lanthanum solution, La concentration 16.5 g/l. Dissolve 12.86 g of lanthanum nitrate hexahydrate, $La(NO_3)_3 \cdot 6H_2O$ in water and make up to 250 ml.

(3) Stock solution, Mn concentration 500 mg/l. Dissolve 1.438 g of potassium permanganate, $KMnO_4$, (standardized according to Appendix A-21), in about 200 ml of water in a beaker. Add carefully, while stirring, 45 ml of concentrated sulphuric acid (96%). Reduce the permanganate with a few drops of hydrogen peroxide (30%) and boil to remove the excess of H_2O_2. Transfer the contents of the beaker quantitatively to a 1000-ml volumetric flask and make up to the mark.

(4) Standard solution, Mn concentration 50 mg/l. Pipette 50.00 ml of the stock solution (5) into a 500-ml volumetric flask and make up to volume with 0.8 M sulphuric acid (1).

Remark:

- Frequently, Mn determinations go along with Zn determinations. In that case, a mixed standard solution may be prepared with 25 mg/l Mn and 20 mg/l Zn (see 2.8 – reagent 3).

Standard series

Pipette 0–1.00–2.00–4.00–6.00–8.00–10.00 ml of the standard solution (4) into 100-ml volumetric flasks and make up to volume with 0.8 M sulphuric acid (1). This standard series has Mn concentrations of 0–0.5–1–2–3–4–5 mg/l.

Procedure

Pipette 5.00 ml of the sample digests, the blank digests and the standard series into test tubes. Add 0.50 ml of lanthanum solution (2) and mix. Nebulize all these solutions into a blue (oxidizing) air-acetylene flame and measure the absorbance at 279.5 nm. Plot a calibration curve and read the manganese concentrations.

Remarks:

- The calibration curve is slightly bent towards the x-axis, so that calculation by means of linear regression is not allowed.
- In a 0.08 M sulphuric acid medium, no interferences were found after spiking the digests with various salts, so as to reach final concentrations of 80 mg/l Al, 500 mg/l Ca, 80 mg/l Fe, 800 mg/l Mg or 16 mg/l Si. However, a concentration of 1000 mg/l Ca results in a precipitate of $CaSO_4$ if the solution has been left standing overnight.

Calculation

The manganese content of the dried plant material, expressed in mg/kg Mn, is calculated by

$$(a - b) \times \frac{V}{w}$$

in which
a = concentration of manganese in the sample digest, in mg/l;
b = concentration of manganese in the blank digest, in mg/l;
V = total volume of digest at the end of the digestion procedure, in ml;
w = weight of plant material sample, in g.

2.5.1 DETERMINATION OF **NITROGEN** BY COLORIMETRY

Field of application

This determination is intended to be carried out on digest 2.0 (H_2SO_4-salicylic acid-H_2O_2, in flasks).

If, for any reason, this determination cannot be carried out, then determination 2.5.2 may be done.

Principle

The determination is based on the Berthelot reaction, in which a phenol derivative (here: salicylate) forms a coloured compound in the presence of ammonia and hypochlorite. In alkaline medium, the indophenol thus formed has a green-blue colour, of which the absorbance is measured at a wavelength of 660 nm. This is a measure for the concentration of ammonium, formed by the nitrogen compounds in the sample.

Remarks:

- To prevent precipitation of hydroxides, EDTA should be added prior to raising the pH.
- The pH should be adjusted with a buffer solution, since the digests contain variable amounts of acid.
- Sodium nitroprusside is a catalyst for the Berthelot reaction.

Apparatus

- Colorimeter.

Reagents

(1) Sodium hydroxide solution, c(NaOH) = 10 mol/l. Dissolve 200 g of sodium hydroxide, NaOH, in about 400 ml water; allow to cool and make up to 500 ml.

(2) Salicylate solution. Dissolve 110 g of salicylic acid, $C_7H_6O_3$, in 105 ml of sodium hydroxide solution (1) and make up directly with water to 250 ml. Prepare just before use.
(3) Buffer solution pH 12.3. Dissolve 26.70 g of disodium hydrogen phosphate dihydrate, $Na_2HPO_4 \cdot 2H_2O$, in some water in a 2-litre volumetric flask. Add 10 ml of the sodium hydroxide solution (1) and dilute to volume with water. Measure the pH and adjust if necessary.
(4) EDTA solution. Dissolve 4 g of disodium ethylene diamine tetra acetate dihydrate, $Na_2EDTA \cdot 2H_2O$, in 100 ml water.
(5) Hypochlorite solution. A stock solution, containing approximately 1 M sodium hypochlorite in 0.1 M NaOH, should be purchased commercially. Dilute 20 ml of this stock solution with water to 100 ml. Prepare fresh daily.
(6) Nitroprusside solution. Dissolve 50 mg of sodium nitroprusside dihydrate, $Na_2[Fe(CN)_5NO] \cdot 2H_2O$, in 100 ml water. Prepare just before use.
(7) Mixed reagent I. Mix 50 ml of the salicylate solution (2) with 100 ml of the nitroprusside solution (6) and 5 ml of the EDTA solution (4).
(8) Mixed reagent II. Mix 200 ml of the buffer solution (3) with 50 ml of the hypochlorite solution (5).
(9) Stock solution, N concentration 2500 mg/l. Dissolve 11.793 g of ammonium sulphate, $(NH_4)_2SO_4$, in water in a 1000-ml volumetric flask and make up to volume with water.

Remark:
- The final solution of sodium hypochlorite should contain 0.7% ± 0.1 % of active chlorine. This will be true for solutions purchased from a supplier of chemicals; if, however, the hypochlorite was bought as bleach in a supermarket, a check on its concentration is recommended.

Standard series

Pipette into 100-ml volumetric flasks, that contain already about 50 ml water, 4.5 ml of concentrated sulphuric acid (96%). Mix and let cool. Then add 0–1.00–2.00–3.00–4.00–5.00–6.00 ml of the stock solution (9) and dilute to volume with water.

Dilute this standard series 1 + 9 (v/v) with water; this should be done together with the diluting of the digests. The diluted standard series has N concentrations of 0–2.5–5 –7.5–10–12.5–15 mg/l.

Procedure

Dilute all digests 1 + 9 (v/v) with water.

Pipette 0.20 ml of the diluted sample digests, the diluted blank digests and the diluted standard series into test tubes. Add 3.0 ml of mixed reagent I (7) and mix. Then add 5.0 ml of mixed reagent II (8) and mix. Allow to stand for at least 2 h. Measure the absorbance in a 1-cm cuvette at a wavelength of 660 nm. Plot a calibration curve, and read the N concentrations.

Remarks:

- The highest standard (15 mg/l N) should give an absorbance reading of approximately 0.68 units.
- In 2 hours the blue colour reaches its maximum intensity; it is stable for at least 10 h.
- The digests obtained according to procedure 2.0 are 0.7–0.9 M in H_2SO_4. Any further dilutions (if the N concentrations would be higher than the highest standard) should be made with zero standard solution.
- The test tubes must be used exclusively for this N determination. They must first be cleaned by taking a blank determination; thereafter they should be cleaned only with water.

Calculation

The nitrogen content of the dried plant material, expressed in mmol/kg N, is calculated by

$$0.714 \times (a - b) \times \frac{V}{w}$$

in which
a = concentration of nitrogen in the diluted sample digest, in mg/l;
b = concentration of N = nitrogen in the diluted blank digest, in mg/l;
V = total volume of digest at the end of the digestion procedure, in ml;
w = weight of plant material sample, in g.

REFERENCES

I. Novozamsky, R. van Eck, J.Ch. van Schouwenburg and I. Walinga. Total nitrogen determination in plant material by means of the indophenol blue method. Neth.J.agric.Sci. *22* (1974) 3–5.

2.5.2 DETERMINATION OF **NITROGEN** BY TITRATION AFTER DISTILLATION

Field of application

This determination is intended to be carried out on digest 2.0 (H_2SO_4-salicylic acid-H_2O_2, in flasks).

Although this determination gives satisfactory results, it is somewhat outdated; therefore, determination 2.5.1 is preferred.

Principle

Ammonia, originating from digestion of nitrogen compounds, is distilled over into boric acid solution. The borate thus formed is titrated with hydrogen iodate to the colour change of an acid-base indicator.

Remarks:

- Since $H(IO_3)_2^-$ is a strong acid, the ammonium borate formed by the distillation is titrated to boric acid; thus, the consumption of $H(IO_3)_2^-$ is directly proportional to the amount of ammonia. As a consequence, neither the volume nor the concentration of the boric acid solution used to capture the ammonia needs to be known accurately.
- Sulphuric acid is often used for the titration. Potassium hydrogen iodate, however, has the advantage of being a primary standard. Yet, the procedure should be checked (see below).
- Although any acid-base indicator can be used, it is known that the sharpest end points are obtained with mixed indicators. Here, a mixture of bromocresol green and methyl red is used.

Apparatus

- Steam generator, electrically heated (see Fig. 3).
- Distillation unit ("Markham still"), micro or semi-micro design (see Fig. 4).
- Piston burette.

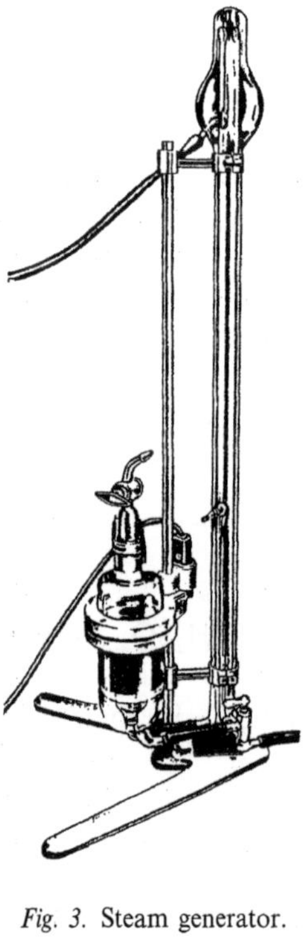

Fig. 3. Steam generator.

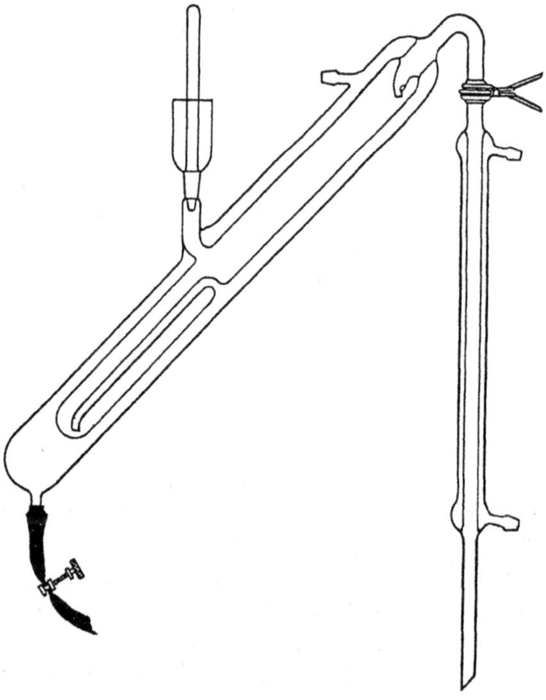

Fig. 4. Distillation unit.

Remarks:

- This steam generator is protected against overloading; the production of steam is adjusted automatically according to the consumption of steam. In the authors' laboratory, a Büchi DG-1500 steam generator is used (manufactured by Büchi, CH-9230 Flawil, Switzerland). This still has the evaporator vessel enclosed in a combined steam jacket and trap: an arrangement which is compact and accelerates distillation. In the authors' laboratory, a Quickfit 46 MC distillation unit is used (manufactured by Corning Ltd., Laboratory Division, Stone, Staffordshire ST15 0BG, England).

Reagents

(1) Sodium hydroxide solution, c(NaOH) = 12.5 mol/l. Dissolve carefully, in portions, 250 g of sodium hydroxide, NaOH, in about 400 ml water. Allow to cool and make up to 500 ml with water.
(2) Boric acid solution. Dissolve 10 g of boric acid, H_3BO_3, in 1 litre of water.
(3) Indicator solution. Dissolve 0.15 g of bromocresol green and 0.1 g of methyl red in 200 ml of ethanol 96% (v/v).
(4) Hydrogen iodate solution, $c(H(IO_3)_2^-)$ = 0.01 mol/l. Dissolve 3.90 g of potassium hydrogen iodate, $KH(IO_3)_2$, in 1000 ml water.
(5) Standard solution, $c(NH_3)$ = 0.0100 mol/l. Dissolve 0.6607 g of ammonium sulphate, $(NH_4)_2SO_4$, in about 400 ml water in a 1000-ml volumetric flask. Add 45 ml of concentrated sulphuric acid (96%), allow to cool and make up to volume.

Standard series

A full standard series is not necessary, but the procedure should be checked by means of standard solution (5) and a zero standard solution, for two reasons. Firstly, the actual procedure may give a result that deviates from the theoretically expected relationship between the amounts of hydrogen periodate and ammonia. Secondly, the optimum distillation time should be established for the apparatus to be used.

Procedure

(a) Standard solution
Pipette 5.00 ml of standard solution (5) into the distillation unit. Add 2.5 ml of sodium hydroxide solution (1) and rinse with some water to a total volume of about 20 ml. Heat the liquid with steam; when the first drop of distillate appears, continue distilling during n minutes (n = 3, 4, 5 or 6), while leading the distillate into 10 ml of boric acid solution (2), to which 6 drops of indicator solution (3) have been added. Take care that the end of the condenser is immersed in the boric acid solution to prevent any loss of ammonia.

Allow the inner side of the condenser tip to be washed by the last portions of distillate by means of lowering – 0.5 min before the distillation is finished – the receiver with boric acid until the end of the condenser is out of the solution.

Titrate the boric acid solution with hydrogen iodate solution (4) until the colour changes from green to rose.

Repeat this procedure with different distillation times. Choose the optimum time and use this when distilling the samples.

Repeat the procedure with a zero standard solution.

Calculate the exact titer of the hydrogen iodate solution, using the results found with the optimum distillation time and correcting for the zero standard, by

$$t_{HIO3} = 0.0500 \times \frac{1}{(V_s - V_z)}$$

in which

t_{HIO3} = titer (concentration) of the hydrogen iodate solution (4), in mol/l;

V_s = volume of hydrogen iodate solution consumed for titrating the standard solution, in ml;

V_z = volume of hydrogen iodate solution consumed for titrating the zero standard solution, in ml.

(b) Sample solution

Pipette 5.00 ml (containing at most 0.1 mmol NH_3) of digest into the distillation unit. Add 2.0 ml of sodium hydroxide solution (1) and rinse with some water to a total volume of about 20 ml. Heat the liquid with steam and distil during the previously established optimum time, in the way described at (a).

Titrate the boric acid solution with standardized hydrogen iodate solution (5) until the colour changes from green to rose.

Repeat the procedure with a blank digest.

Calculation

The total nitrogen content of the dried plant material, expressed in mmol/kg N, is calculated by

$$200 \times t \times (a - b) \times \frac{V}{w}$$

in which

t = concentration of the hydrogen iodate solution, in mol/l;

a = volume of hydrogen iodate solution consumed by the sample digest, in ml;

b = volume of hydrogen iodate solution consumed by the blank digest, in ml;

V = total volume of digest at the end of the digestion procedure, in ml;

w = weight of plant material sample, in g.

2.6 DETERMINATION OF **SODIUM** BY FLAME AES

Field of application

This determination is intended to be carried out on digest 2.0 (H_2SO_4-salicylic acid-H_2O_2, in flasks).

The determination may be done in combination with K (2.2).

Principle

The sample is nebulized into an air-propane flame, where it is vaporized; sodium compounds are atomized and the sodium atoms thus formed emit radiation of which the intensity is measured at a wavelength of 589.0 nm.

Remarks:

- To prevent ionisation interferences, a cesium salt is added to act as an ionization buffer.
- Since the sodium atomic emission line finds itself in the wavelength region of a calcium oxide emission band, too high results will be found if calcium is present in the digest; however, since calcium in the sulphuric acid digests will form undissociated $CaSO_4$ in the flame, the interference of CaO is negligible.

Apparatus

- Flame emission spectrometer.

Reagents

(1) Cesium solution, Cs concentration 0.86 g/l. Dissolve 1.1 g of cesium chloride, CsCl, "pro analysi", in 1 litre of water.

(2) Stock solution, Na concentration 1250 mg/l. Dissolve 3.178 g of sodium chloride, NaCl, pretreated according to Appendix A-4, in some water in a 1000-ml volumetric flask and make up to the mark with water.

Remarks:

- The cesium chloride should be of the highest analytical grade ("pro analysi"), because a lower quality (e.g. "reinst" = "most pure") contains much more Na.
- Frequently, Na determinations go along with K determinations. In that case, a mixed stock solution may be prepared with 1250 mg/l Na and 4000 mg/l K (see 2.2 – reagent 2).

Standard series

Pipette 0–2.00–4.00–6.00–8.00–10.00–12.00 ml of the stock solution (2) into 100-ml volumetric flasks, to which already 50 ml water and 4.5 ml of concentrated sulphuric acid (96%) have been added in advance. Make up to the mark with water. Dilute this standard series 1 + 9 (v/v) with cesium solution (1); this should be done together with the diluting of the sample digests. This diluted standard series has Na concentrations of 0–2.5–5–7.5–10–12.5–15 mg/l.

Procedure

Dilute all digests 1 + 9 (v/v) with the cesium solution (1). Measure the emission intensity of the diluted sample digests, the diluted blank digests and the diluted standard series at a wavelength of 589.0 nm, using an air-propane flame. Plot a calibration curve and read the sodium concentrations.

Remarks:

- The calibration curve should be nearly linear.
- An air-acetylene flame can be used also, since the Cs concentration is high enough to counteract the greater tendency to ionisation.
- Instead of indicating a wavelength, a simple emission spectrometer ("flame photometer") may only be supplied with a so-called Na filter. Make sure that this is an interference filter, since glass filters are not selective enough.

Calculation

The sodium content of the dried plant material, expressed in mmol/kg Na, is calculated by

$$0.435 \times (a - b) \times \frac{V}{w}$$

in which

a = concentration of sodium in the diluted sample digest, in mg/l;
b = concentration of sodium in the diluted blank digest, in mg/l;
V = total volume of digest at the end of the digestion procedure, in ml;
w = weight of plant material sample, in g.

2.7 DETERMINATION OF **PHOSPHORUS** BY COLORIMETRY

Field of application

This determination is intended to be carried out on digest 2.0 (H_2SO_4-salicylic acid-H_2O_2, in flasks).

Principle

In an acidic medium, orthophosphates form a yellow-coloured complex with molybdate ions. After addition of ascorbic acid and Sb, a blue-coloured phosphomolybdenum complex is formed. Its absorbance is measured at a wavelength of 880 nm.

Remarks:

- As^{5+} and Si^{4+} form similarly coloured complexes under these conditions; As^{5+} interference, if any, may be suppressed by reduction of As^{5+} to As^{3+} (see remark with the procedure). Si^{4+} does not normally interfere with the digests in question, because these have been decanted.
- The blue colour varies with the redox conditions of the medium and with pH.

- Sb is added to complete the reduction of the yellow phosphomolybdenum complex to the blue phosphomolybdenum complex. Moreover, Sb enhances the intensity of the blue colour and allows a most sensitive measurement at 880 nm.

Apparatus

- Colorimeter, allowing measurements at 880 nm.

Reagents

(1) Sulphuric acid, $c(H_2SO_4) = 2.5$ mol/l. Add carefully, while swirling, 140 ml of concentrated sulphuric acid (96%) to about 400 ml water. Allow to cool and make up to 1 litre.
(2) Sulphuric acid, $c(H_2SO_4) = 0.7$ mol/l. Add carefully, while swirling, 40 ml of concentrated sulphuric acid (96%) to about 400 ml water. Allow to cool and make up to 1 litre.
(3) Molybdate solution. Dissolve 4 g of ammonium molybdate, $(NH_4)_6Mo_7O_{24} \cdot 4H_2O$ in 100 ml water. Store in a hard glass vessel.
(4) Ascorbic acid solution. Dissolve 1.76 g of ascorbic acid, $C_6H_8O_6$, in 100 ml water. Prepare fresh daily.
(5) Antimonyl tartrate solution. Dissolve 0.14 g of potassium antimonyl tartrate, $KSbC_4H_4O_7$, in 50 ml water.
(6) Reduction mixture. Add 2 g of sodium metabisulphite, $Na_2S_2O_5$, and 0.2 g of sodium thiosulphate, $Na_2S_2O_3$, to about 200 ml water. Introduce a magnetic bar and stir until all crystals have dissolved. Add slowly, while still stirring, 10 ml of 2.5 M sulphuric acid (1); then make up to 250 ml. Prepare just before use.
(7) Mixed reagent. Add 15 ml of molybdate solution (3) to 50 ml of 2.5 M sulphuric acid (1) and mix. Next add 30 ml of ascorbic acid solution (4) and mix; then add 5 ml of antimonyl tartrate solution (5) and mix. Prepare fresh daily. See remark with procedure.
(8) Diluted mixed reagent. Mix 80 ml of the mixed reagent (7) with 300 ml water.
(9) Stock solution, P concentration 500 mg/l. Dissolve 2.197 g of potassium dihydrogen phosphate, KH_2PO_4, in about 400 ml water in a 1000-ml volumetric flask. Add carefully, while swirling, 40 ml of concentrated sulphuric acid (96%). Allow to cool and make up to the mark.

Remark:
- The addition of sulphuric acid (1) to the reduction mixture (6) should be done slowly and while swirling, because otherwise the thiosulphate may decompose and form a precipitate of elemental S.

Standard series

Pipette 0–2.00–4.00–6.00–8.00–10.00 ml of the stock solution (9) into 100-ml volumetric flasks. Dilute to volume with 0.7 M sulphuric acid (2).

Dilute this standard series 1 + 9 (v/v) with water; this should be done together with the diluting of the digests. The diluted standard series has P concentrations of 0–1–2–3–4–5 mg/l.

Procedure

Dilute all digests 1 + 9 (v/v) with water.

Pipette 1.0 ml of the diluted sample digests, the diluted blank digests and the diluted standard series into test tubes. Add 3.8 ml of the diluted mixed reagent (8) and mix. Allow to stand for 1 hour. Measure the absorbance in a 1-cm cuvette at a wavelength of 880 nm. Plot a calibration curve and read the phosphorus concentrations.

Remarks:

- After having reached its maximum intensity, the colour is stable for at least 5 h.
- The absorbance can also be measured at 720 nm or 600 nm. The measurement is then somewhat less sensitive.
- Only part of the Mo^{6+} will be reduced. Hence, samples that have high phosphate concentrations cannot be diluted once that the blue colour has been formed.
- When As^{5+} is present, 3 ml of the reduction mixture (6) should be added to all solutions; after 15 minutes waiting, 0.8 ml of the *undiluted* mixed reagent (7) is added.

Calculation

The phosphorus content of the dried plant material, expressed in mmol/kg P, is calculated by

$$0.323 \times (a - b) \times \frac{V}{w}$$

in which

a = concentration of phosphorus in the diluted sample digest, in mg/l;
b = concentration of phosphorus in the diluted blank digest, in mg/l;
V = total volume of digest at the end of the digestion procedure, in ml;
w = weight of plant material sample, in g.

REFERENCES

J.Ch. van Schouwenburg and I. Walinga. The rapid determination of phosphorus in presence of arsenic, silicon and germanium. Anal. Chim. Acta *37* (1967) 271–274.

2.8 DETERMINATION OF **ZINC** BY FLAME AAS

Field of application

This determination is intended to be carried out on digest 2.0 (H_2SO_4-salicylic acid-H_2O_2, in flasks).

The zinc content of the dried plant material should be higher than 15 mg/kg.

The determination may be done in combination with Mn (2.4).

Principle

The sample is nebulized into an air-acetylene flame, where it is vaporized; zinc compounds are atomized and the zinc atoms thus formed absorb radiation from a hollow-cathode lamp. The absorbance is measured at a wavelength of 213.9 nm.

Apparatus

- Atomic absorption spectrophotometer.

Reagents

(1) Sulphuric acid, c(H_2SO_4) = 0.8 mol/l. Add carefully, while swirling, 45 ml of concentrated sulphuric acid (96%) to about 400 ml water. Allow to cool and make up to 1 litre.

(2) Stock solution, Zn concentration 400 mg/l. Dissolve 1.759 g of zinc sulphate heptahydrate, $ZnSO_4 \cdot 7H_2O$ (standardized according to Appendix A-24) in about 600 ml water in a 1000-ml volumetric flask. Add 45 ml of concentrated sulphuric acid (96%), allow to cool and make up to volume.

(3) Standard solution, Zn concentration 20 mg/l. Pipette 25.0 ml of the stock solution (2) into a 500-ml volumetric flask and make up to volume with 0.8 M sulphuric acid (1).

Remark:

- Frequently, Zn determinations go along with Mn determinations. In that case, a mixed standard solution may be prepared with 20 mg/l Zn and 50 mg/l Mn (see 2.4 – reagent 3).

Standard series

Pipette 0–1.00–2.00–4.00–6.00–8.00–10.00 ml of the standard solution (3) into 100-ml volumetric flasks and make up to volume with 0.8 M sulphuric acid (1). This standard series has Zn concentrations of 0–0.2–0.4–0.8–1.2–1.6–2.0 mg/l.

Procedure

Nebulize the sample digest, the blank digests and the standard series into a just blue (stoichiometric) air-acetylene flame and measure the absorbance at 213.9 nm. Plot a calibration curve and read the zinc concentrations.

Remarks:

- The calibration curve is slightly bent towards the x-axis, so that calculation by means of linear regression is not allowed.
- Although the chosen wavelength is short, no background correction has been found necessary when using 0.08 M sulphuric acid. For other media, background correction may be required.
- No interferences were observed after spiking the digests with several salts, so as to reach final concentrations of 100 mg/l Mg, 30 mg/l Fe, 20 mg/l Al + 400 mg/l PO_4, or 30 mg/l Fe + 20 mg/l Al + 400 mg/l PO_4.
- Do not use rubber stoppers, since these will release zinc.

Calculation

The zinc content of the dried plant material, expressed in mg/kg Zn, is calculated by

$$(a - b) \times \frac{V}{w}$$

in which

a = concentration of zinc in the sample digest, in mg/l;
b = concentration of zinc in the blank digest, in mg/l;
V = total volume of digest at the end of the digestion procedure, in ml;
w = weight of plant material sample, in g.

3 Digestion with HNO_3-$HClO_4$-H_2SO_4 and determination of Al, Cd, Cu, Fe, Mn, Pb, Zn

3.0 DIGESTION WITH HNO_3-$HClO_4$-H_2SO_4

Field of application

In this digestion a fairly large amount of plant material is weighed out, which allows the determination of the trace elements Al, Cd, Cu, Fe, Mn, Pb and Zn.

Principle

Easily oxidizable organic matter is attacked by HNO_3 at relatively low temperature. After the excess HNO_3 is distilled off, the remaining organic compounds are oxidized by $HClO_4$ at elevated temperature.

Remarks:

- Sulphuric acid is present to "dilute" the perchloric acid in the final stage so as to prevent the risk of explosions.
- See Appendix A-6 for safety precautions on perchloric acid.

- This digestion method is also known as the “Schaumlöffel” method (see literature reference).

Apparatus

- Heating mantles or blocks for 100-ml flasks.
- Metal weighing funnel (see Fig.1).
- Digestion apparatus adapted from Chat (see Figs. 2 and 3).

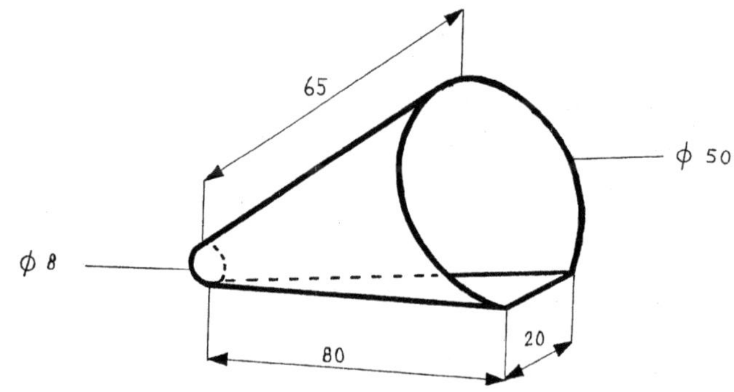

Fig. 1. Weighing funnel (measures in mm).

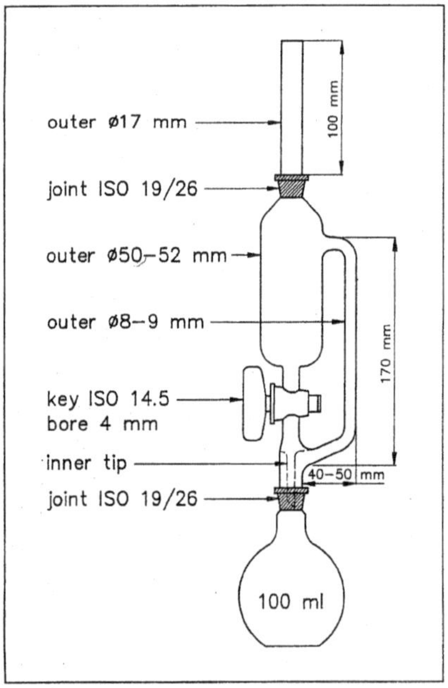

Fig. 2. Digestion apparatus.

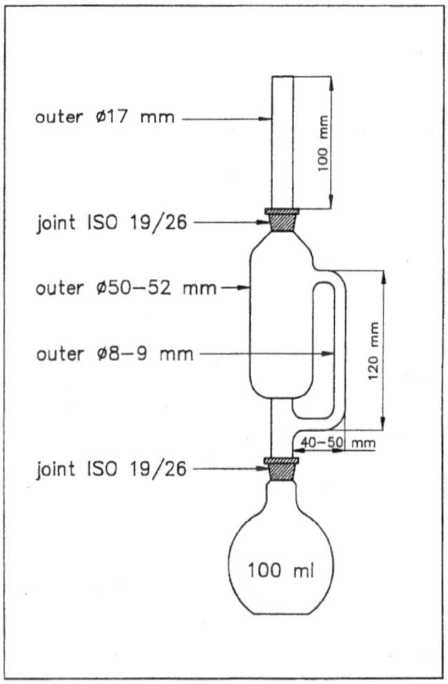

Fig. 3. Digestion apparatus as in use in the authors' laboratory.

Remarks:

- Dried plant material may easily stick to glass when the relative humidity of the air is low. These home-made weighing funnels (stainless steel or aluminium) do not show this effect.
- The Chat digestion apparatus as shown in Fig. 2 is commercially available. The dimensions of similar glassware from other suppliers may be slightly different, without affecting the digestion process. The stopcock is not necessary, so that a similar condenser (either purchased commercially or home-made) can be used as well. In the authors' laboratory, actually, custom-made digestion apparatus as shown in Fig.3 are in use.

Reagents

(1) Nitric acid, 65% (w/w), $c(HNO_3)$ = 14.4 mol/l (d = 1.40 g/ml).
(2) Perchloric acid, 70% (w/w), $c(HClO_4)$ = 11.6 mol/l (d = 1.67 g/ml).
(3) Sulphuric acid, 96% (w/w), $c(H_2SO_4)$ = 18 mol/l (d = 1.84 g/ml).
(4) Digestion mixture. Mix 400 ml of nitric acid (1) with 40 ml of perchloric acid (2) and add 10 ml of sulphuric acid (3).
(5) Sodium nitrite solution, 5 g/l. Dissolve 0.5 g of $NaNO_2$ in 100 ml of water.
(6) Carborundum beads.

Procedure

Weigh, to the nearest 0.001 g, approximately 2 g of the dried plant material sample in a metal weighing funnel and transfer the sample to the flat-bottomed 100-ml flask of the digestion apparatus. Add 20 ml of the digestion mixture (4) and 4 carborundum beads (6). Wet the plant material by swirling the flask. Moisten the lower cone of the condenser with water and place it on the flask. Moisten the cone of the upper glass tube with water and place it on top of the condenser. The stopcock should be closed. Fix the whole digestion apparatus upright and allow to stand overnight at room temperature. Prepare also two blank digestions.

Heat moderately (about 170 °C) for at least 40 min until most of the nitric acid has distilled off. Then raise the temperature stepwise, so that the remainder of nitric acid, together with some water, distills over. During this process the contents of the flask gradually turns black. When the distillation stops, the temperature will increase until the azeotropic boiling point of perchloric acid (205 °C) is reached. The now concentrated perchloric acid attacks remaining organic material with a violent oxidation reaction, in which dense white fumes are developed. Adjust the heat supply in such a way that these perchloric acid vapours condense halfway the side-arm. When the digest has turned colourless, or slightly coloured, continue digestion for 1 h more. Then cool a little, add about 20 ml water and 2 ml of sodium nitrite solution (5). Boil for 10 min, cool, discard the contents of the condenser, and rinse the side-arm of the condenser, thereby collecting all liquid in the flat-bottomed digestion flask. Transfer the contents of this flask to a 100-ml volumetric flask; wash and collect these washings in the same volumetric flask and make up to the mark. Mix and filter over coarse paper into a 100-ml erlenmeyer flask.

Remarks:

- The plant material has to stand overnight with the digestion mixture to prevent excessive foaming.
- The first stage of the digestion should take at least 40 min to allow the nitric acid to destroy all the easily oxidizable material. This prevents any explosions due to too rapid reaction by perchloric acid.
- The hot plates in the authors' laboratory are set at position 4–5 for 170 °C and at 5.5–6.5 for 205 °C.
- Sodium nitrite reduces insoluble higher oxides of manganese which might have been formed.
- Boiling is required to dissolve salts after the digestion.
- The cones and stopcock must never be greased, but should be moistened with water.
- The sulphate concentration in the blank digest is 0.08 mol/l.

REFERENCES

E. Schaumlöffel. Über die colorimetrische Bestimmung der Mikronährstoffe Kupfer, Zink, Kobalt, Mangan, Eisen und Molybdän aus einer Aschenlösung durch fraktionierte Extraktion. Landwirtschaftliche Forschung *13* (1960) 278–286. (in German)

G. Chat. Nouvelle méthode de minéralisation des végétaux pour les analyses chimiques. Académie d'Agriculture de France, procès-verbal de la séance du 9 Novembre 1966, p. 1087 à 1093. (in French)

3.1 DETERMINATION OF **ALUMINIUM** BY COLORIMETRY

Field of application

This determination is intended to be carried out on digest 3.0 (HNO_3-$HClO_4$-H_2SO4).

The aluminium content of the dried plant material should be higher than 25 mg/kg.

Principle

Aluminium ions form – in weakly acid medium and in the presence of a polycyclic ketoamine – an extremely stable, red-coloured complex with Eriochrome Cyanine R.

Remark:

- This reagent is specific for Al; the only possible interference comes from Fe(III), but only at concentrations which are normally not found in plant material. Any Fe interference is prevented by addition of ascorbic acid.

Apparatus

– Colorimeter.

Reagents

(1) Polycyclic ketoamine, a heterocyclic fatty acid amine, to be purchased from Amchem Products, Inc., Ambler, Pa, USA, or from N.V. Mavom, Handelsweg 6, Alphen a/d Rijn, The Netherlands.
(2) Sulphuric acid, c(H_2SO_4) = 0.08 mol/l. Add 4.5 ml of concentrated sulphuric acid (96%) to about 400 ml water and make up to 1 litre.
(3) Ammonium nitrate solution. Dissolve 6.25 g of ammonium nitrate,NH_4NO_3, in 50 ml water.
(4) Sodium chloride solution. Dissolve 6.25 g of sodium chloride, NaCl, in 50 ml water.
(5) Eriochrome Cyanine R solution. Dissolve 250 mg of Eriochrome Cyanine R (C.I. nr. 43820) in 50 ml water. Add 50 ml of ammonium nitrate solution (3) and mix. Add 50 ml of sodium chloride solution (4) and mix. Add 2 ml of concentrated nitric acid (65%) and mix. Make up to 250 ml with water.
(6) Diluted Eriochrome Cyanine R solution. Add 60 ml water to 40 ml of the Eriochrome Cyanine R solution (5).
(7) Ascorbic acid solution. Dissolve 1 g of ascorbic acid, $C_6H_8O_6$, in 50 ml water. Prepare fresh daily.
(8) Sulphite solution. Dissolve 1.25 g of sodium sulphite, Na_2SO_3, (or 2.5 g of sodium sulphite heptahydrate) in 50 ml water. Prepare fresh daily.
(9) Polycyclic ketoamine solution. Dissolve 0.7 g of polycyclic ketoamine (1) in 100 ml water. Prepare fresh daily.
(10) Acetate solution. Dissolve 20 g of sodium acetate trihydrate, $CH_3COONa \cdot 3H_2O$, in water and make up to 50 ml.
(11) Mixed reagent. Mix 50 ml of sulphite solution (8) with 50 ml of polycyclic ketoamine solution (9) and 50 ml of acetate solution (10). Add 225 ml water and mix. Prepare fresh daily. Do not use rubber stoppers.
(12) Stock solution, Al concentration 50 mg/l. Dissolve 0.8791 g of potassium aluminium sulphate dodecahydrate, $KAl(SO_4)_2 \cdot 12H_2O$, (standardized according to App. A-15) in some water in a 1000-ml volumetric flask, add 4.5 ml of concentrated sulphuric acid (96%) and make up to the mark.

Standard series

Pipette 0–2.00–4.00–6.00–8.00–10.00 ml of the stock solution (12) into 100-ml volumetric flasks. Make up to the mark with 0.08 M sulphuric acid (2). Transfer the solutions to polythene bottles. This standard series has Al concentrations of 0–1–2–3–4–5 mg/l.

Procedure

Pipette 0.20 ml of the sample digests, the blank digests and the standard series into test tubes. Add successively

- 1.0 ml ascorbic acid solution (7); mix and wait for 5 min;
- 1.0 ml diluted Eriochrome Cyanine R solution (6); mix and wait for 5 min;
- 3.0 ml mixed reagent (11); mix and wait for 1 h.

Measure the absorbance in a 1-cm cuvette at a wavelength of 595 nm. Plot a calibration curve and read the aluminium concentrations.

Remarks:

- The colour of the measuring solution is stable for 2 h.
- The calibration curve should be linear, but may flatten at the lowest concentrations.
- All glassware must be pretreated with "chromic acid", to prevent Al desorption and to cleanse it from polycyclic ketoamine. Do not use a detergent, since this is contaminated by aluminium.
- The pH of the final solution should lie between 5.5 and 5.1. The absorbance will increase with pH by 0.02 units per pH unit.
- No interferences were found with 10 mg/l Ni, 50 mg/l Cu, 100 mg/l Mn and 10 mg/l Co at an aluminium level of 2 mg/l. A concentration of 200 mg/l Fe raises the apparent aluminium concentration by 1%.

Calculation

The aluminium content of the dried plant material, expressed in mg/kg Al, is calculated by

$$(a - b) \times \frac{V}{w}$$

in which

a = concentration of aluminium in the sample digest, in mg/l;
b = concentration of aluminium in the blank digest, in mg/l;
V = total volume of digest at the end of the digestion procedure, in ml;
w = weight of plant material sample, in g.

REFERENCES

U.T. Hill. New direct spectrophotometric determination of aluminium in steel, spelter, and iron ores. Anal. Chem. *38* (1966) 654–656.

3.2.1 DETERMINATION OF **CADMIUM** BY FLAME AAS

Field of application

This determination is intended to be carried out on digest 3.0 (HNO_3-$HClO_4$-H_2SO_4).

The cadmium content of the dried plant material should be higher than 5 mg/kg.

Samples with lower Cd levels should be determined by method 3.2.2; since this method is more cumbersome, it is recommended to apply method 3.2.1 first to all samples as a screening.

Principle

The sample is nebulized into an air-acetylene flame, where it is vaporized; cadmium compounds are atomized and the cadmium atoms thus formed absorb radiation from a hollow-cathode lamp. The absorption is measured at a wavelength of 228.8 nm.

Remarks:

- The plant material digest appears to create a continuous background absorption, so that background correction is necessary.
- Extremely high Fe levels (> 5 g/kg dry material) will cause noticeable spectral interference because of a structured background. When using deuterium background correction, this may lead to too low analytical results (see literature reference).

Apparatus

- Atomic absorption spectrophotometer, with a device for correcting or measuring background absorption.

Reagents

(1) Sulphuric acid, $c(H_2SO_4) = 0.8$ mol/l. Add carefully, while swirling, 45 ml of concentrated sulphuric acid (96%) to about 400 ml water. Allow to cool and make up to 1 litre.

(2) Stock solution, Cd concentration 1000 mg/l. Dissolve 2.744 g of cadmium nitrate tetrahydrate, $Cd(NO_3)_2 \cdot 4H_2O$ (standardized according to App. A-16) in some water in a 1000-ml volumetric flask. Add 4.5 ml of concentrated sulphuric acid (96%) and make up to the mark.

(3) Standard solution, Cd concentration 10 mg/l. Pipette 1.00 ml of the stock solution (2) into a 100-ml volumetric flask and make up to the mark.

Remarks:

- Reagents and solutions must be prepared with *freshly* delivered water, since polythene and other plastic vessels may release cadmium on standing. The use of extra purified water is not necessary.

- Frequently, Cd determinations go along with Pb determinations. However, no mixed standard solution of the required concentration can be prepared in 0.08 M sulphuric acid, because a precipitate of $PbSO_4$ would be formed. Proceed as described under Standard series.

Standard series

Pipette 0–0.50–1.0–2.0–3.0–4.0 ml of the standard solution (3) into 100-ml volumetric flasks. Add 70 ml of water to each flask. Then add 10 ml of 0.8 M sulphuric acid (1); mix well. (If Pb is also to be measured, the prescribed amounts of standard solution (see 3.6.1) should now be added.) Make up to volume. This standard series has Cd concentrations of 0–0.05–0.1–0.2–0.3–0.4 mg/l.

Remark:

- By first adding the bulk of water, one makes sure that the solubility product of $PbSO_4$ is not surpassed in the sulphuric acid standards.

Procedure

Nebulize the sample digests, the blank digests and the standard series into a just blue (stoichiometric) air-acetylene flame and measure the absorbance at 228.8 nm. plot a calibration curve and read the cadmium concentrations.

Remarks:

- The calibration curve is slightly bent towards the x-axis.
- Background correction should be applied.
- Samples with very high Fe contents will give erroneous results. The iron should be removed (e.g. by extraction with acetylacetone) or standard addition should be applied; alternatively a special background correction system, e.g. Smith-Hieftje, should be used.

Calculation

The cadmium content of the dried plant material, expressed in mg/kg Cd, is calculated by

$$(a - b) \times \frac{V}{w}$$

in which

a = concentration of cadmium in the sample digest, in mg/l;
b = concentration of cadmium in the blank digest, in mg/l;
V = total volume of digest at the end of the digestion procedure, in ml;
w = weight of plant material sample, in g.

REFERENCES

J.J. van der Lee, E. Temminghoff, V.J.G. Houba and I. Novozamsky. Background corrections in the determination of Cd and Pb by flame AAS in plant and soil samples with high Fe levels. Appl. Spectrosc. *41* (1987) 388–390.

3.2.2 DETERMINATION OF **CADMIUM** BY FURNACE AAS

Field of application

This determination is intended to be carried out on digest 3.0 (HNO_3-$HClO_4$-H_2SO_4).

It is intended for samples that are very low in Cd (< 5 mg/kg); the cadmium content should, however, be higher than 25 μg/kg.

Principle

The sample is brought into a graphite furnace, where it is subsequently dried, ashed and vaporized by electrical heating. The cadmium atoms thus formed absorb radiation from a hollow-cathode lamp. The absorbance is measured at a wavelength of 228.8 nm while applying Zeeman background correction.

Remark:

- When using Zeeman background correction, no extraction with an organic sequestering agent into xylene is needed to eliminate matrix interferences.

Apparatus

- Atomic absorption spectrophotometer, with electrothermal atomization device (graphite furnace) and with a device for correcting or measuring background absorption, preferably a Zeeman background correction system.

Reagents

(1) Sulphuric acid, $c(H_2SO_4)$ = 0.8 mol/l. Add carefully, while swirling, 45 ml of concentrated sulphuric acid (96%) to about 400 ml water. Allow to cool and make up to 1 litre.

(2) Sulphuric acid, $c(H_2SO_4)$ = 0.08 mol/l. Dilute the 0.8 M sulphuric acid (1) 1 + 9 (V/V).

(3) 1-Butanol (= n-butylalcohol).

(4) Stock solution, Cd concentration 1000 mg/l. Dissolve 2.7442 g of cadmium nitrate tetrahydrate, $Cd(NO_3)_2 \cdot 4H_2O$, (standardized according to App. A-16) in some water in a 1000-ml volumetric flask. Add 4.5 ml of concentrated sulphuric acid (96%), and make up to the mark.

(5) Standard solution, Cd concentration 10 mg/l. Pipette 1.00 ml of the stock solution (4) into a 100-ml volumetric flask; make up to the mark with 0.08 M sulphuric acid (2).
(6) Diluted standard solution, Cd concentration 50 µg/l. Pipette 5.00 ml of the standard solution (5) into a 1000-ml volumetric flask. Add 100 ml of 0.8 M sulphuric acid (1) and make up to volume.

Remarks:

- Reagents and solutions must be prepared with *freshly* delivered water, since polythene and other plastic vessels may release cadmium on standing. The use of extra purified water is not necessary.
- Frequently, Cd determinations go along with Pb determinations. However, no mixed standard solution of the required concentration can be prepared in 0.08 M sulphuric acid, because a precipitate of PbSO4 would be formed. Proceed as described under Standard series.

Standard series

Pipette 0–0.50–1.00–1.50–2.00 ml of the diluted standard solution (6) into 100-ml erlenmeyer flasks. If Pb is also to be measured, pipette 0–0.50–1.00–1.50–2.00 ml of a 1000 µg/l Pb solution (see 3.6.2) into the same erlenmeyer flasks. Dilute to about 30 ml with 0.08 M sulphuric acid (2); mix immediately after addition of acid.

Procedure

Transfer (part of) the digests and the standard series to polythene cups that fit in the automatic sampler of the atomic absorption spectrophotometer. Add 0.20 ml of 1-butanol (3) and mix thoroughly with an electric mini-stirrer. Heat the extract in a graphite furnace according to an appropriate time-temperature programme (see remark). Measure the absorbance at 228.8 nm in the atomization phase, in the cleaning phase, and at room temperature at the end of the cycle; use Zeeman background correction. Plot a calibration curve and read the cadmium concentrations.

Remarks:

- When a sample gives a higher absorbance than the highest standard, it should be diluted with the zero standard solution (dilution factor = f); this can be programmed in the automatic sampler.
- The calibration curve is slightly bending towards the x-axis.
- Samples with high Fe content may give too low results when applying a deuterium background correction system.
- The absorbance reading in the cleaning phase is meant as a check on memory effects. If these are observed, the results of the last and the second last sample should be discarded. With the programme used in our laboratory (see remark), memory effects are only observed if something is out of order.
- After completion of each firing, the absorbance is read at room temperature as an indication of baseline drift. This value is used as a correction on the reading in the atomization phase.
- Every sample should be measured at least three times, since the precision of this technique is low.

- The wash solution of the automatic sampler contains 2-propanol (2% in water) in order to lower its surface tension.
- The measurements can be performed with any ETA-AAS system. The present method was worked out using a Varian SpectrAA 300 system. The optimum programme – after 30 min warming up – is as follows:

phase	I (dry)	II (char)	III (atomize)	IV (clean)	"V" (base line)
temperature (°C)	160	450	2200	2500	ambient
rate	1	0	1	0	—
hold time (s)	30	45	10	3	—
gas stop (s)	—	—	10	—	—
read	—	—	yes	yes	yes

- When using other instruments, the optimum temperature values to be set may differ from the values given above, because the temperatures mentioned here are instrument settings instead of real temperature values.

Calculation

The cadmium content of the dried plant material, expressed in µg/kg Cd, is calculated by

$$\frac{(a - b) \times f \times V}{w}$$

in which
a = concentration of cadmium in the sample digest, in µg/l;
b = concentration of cadmium in the blank digest, in µg/l;
f = dilution factor for the measuring solution;
V = total volume of digest at the end of the digestion procedure, in ml;
w = weight of plant material sample, in g.

3.2.3 DETERMINATION OF **CADMIUM** BY FURNACE AAS AFTER LIQUID-LIQUID EXTRACTION

Field of application

This determination is intended to be carried out on digest 3.0 (HNO_3-$HClO_4$-H_2SO_4).

It is intended for samples that are very low in Cd (< 5 mg/kg); the cadmium content should, however, be higher than 25 µg/kg.

Method 3.2.3 should only be applied when no Zeeman background correction system is available.

Principle

Cadmium ions are sequestered by a carbamate, and the sample is extracted into xylene. The extract is subsequently dried, ashed and vaporized by electrical heating in a graphite furnace. The cadmium atoms thus formed absorb radiation from a hollow-cathode lamp. The absorbance is measured at a wavelength of 228.8 nm.

Remarks:

- The extraction serves two purposes at a time: a large part of the matrix interferences is eliminated, and the cadmium is more concentrated.
- Since not all interferences will have disappeared, background correction is necessary.

Apparatus

- Erlenmeyer flasks with ground glass necks and PTFE stoppers.
- Linear shaker.
- Water repellent filter paper.
- Polythene cups.
- Atomic absorption spectrophotometer, with electrothermal atomization device (graphite furnace) and with a device for correcting or measuring background absorption.

Reagents

(1) Sulphuric acid, $c(H_2SO_4) = 0.8$ mol/l. Add carefully, while swirling, 45 ml of concentrated sulphuric acid (96%) to about 400 ml water. Allow to cool and make up to 1 litre.
(2) Sulphuric acid, $c(H_2SO_4) = 0.08$ mol/l. Dilute the 0.8 M sulphuric acid (1) 1 + 9 (v/v).
(3) DDDTC solution. Dissolve 5 g of diethylammonium-N,N-diethyldithiocarbamate, $C_9H_{22}N_2S_2$, in 100 ml xylene. Stir with a magnetic stirrer until all crystals are dissolved; this will take about 45 minutes. The solution can be kept overnight at room temperature.
(4) 1-Butanol (= n-butylalcohol).
(5) Stock solution, Cd concentration 1000 mg/l. Dissolve 2.7442 g of cadmium nitrate tetrahydrate, $Cd(NO_3)_2 \cdot 4H_2O$, (standardized according to App. A-16) in some water in a 1000-ml volumetric flask. Add 4.5 ml of concentrated sulphuric acid (96%), and make up to the mark.
(6) Standard solution, Cd concentration 10 mg/l. Pipette 1.00 ml of the stock solution (5) into a 100-ml volumetric flask; make up to the mark with 0.08 M sulphuric acid (2).
(7) Diluted standard solution, Cd concentration 50 μg/l. Pipette 5.00 ml of the standard solution (6) into a 1000-ml volumetric flask. Add 100 ml of 0.8 M sulphuric acid (1) and make up to volume.

Remarks:

- Reagents and solutions must be prepared with *freshly* delivered water, since polythene and other plastic vessels may release cadmium on standing. The use of extra purified water is not necessary.
- Frequently, Cd determinations go along with Pb determinations. However, no mixed standard solution of the required concentration can be prepared in 0.08 M sulphuric acid, because a precipitate of $PbSO_4$ would be formed. Proceed as described under Standard series.
- The solid DDDTC should be stored in a refrigerator in order to prevent deterioration.

Standard series

Pipette 0–0.50–1.00–1.50–2.00 ml of the diluted standard solution (7) into 100-ml erlenmeyer flasks. If Pb is also to be measured, pipette 0–0.50–1.00–1.50–2.00 ml of a 1000 µg/l Pb solution (see 3.6.3) into the same erlenmeyer flasks. Dilute to about 30 ml with 0.08 M sulphuric acid (2); mix immediately after addition of acid.

Procedure

(a) Pipetting of digests

Pipette D ml (D varying between 1 and 30 ml, according to the expected cadmium concentrations) of the sample digests into 100-ml erlenmeyer flasks; pipette also E ml (E = highest D) of the blank digests into 100-ml erlenmeyer flasks. Dilute to 30 ml with 0.08 M sulphuric acid (2).

(b) Extraction

Add both to standard series, sample digests and blank digests 10.00 ml of the DDDTC solution (3) and close the erlenmeyer flasks with a PTFE stopper. Shake mechanically (see Fig. 4) during 4 min (stroke length 9 cm, 150

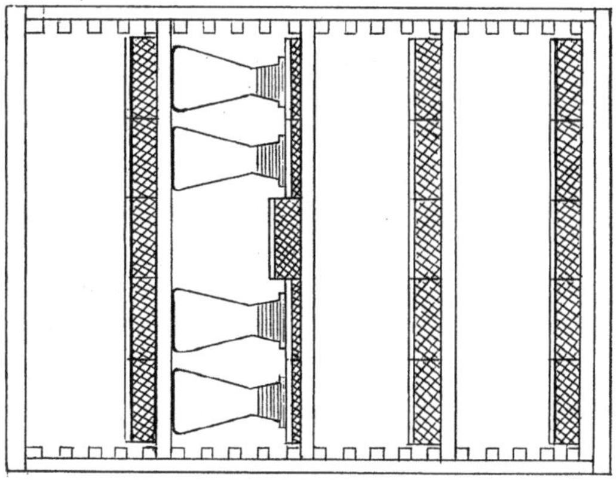

Fig. 4. Rack & erlenmeyer flasks (top view).

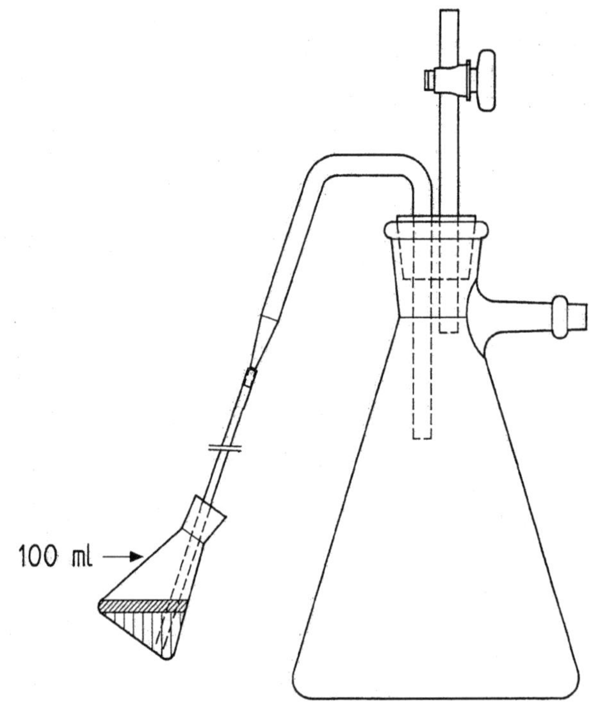

Fig. 5. Sucking off the (lower) water phase.

oscillations/min). Suck off the water phase (see Fig. 5) to such an extent that the remaining water will not occupy more than half of the filter area, and filter over water repellent filter paper.

Shake an extra amount of five portions of the zero standard for future dilution purposes; mix the five extracts to make a uniform zero standard (pool).

After the extraction, the standard series has cadmium concentrations of 0–2.5–5–7.5–10 μg/l. (If lead was also put in, its concentrations are 0–50–100–150–200 μg/l).

(c) Measurement

Pipette 1.00 ml of the extracts obtained at (b) into polythene cups that fit in the automatic sampler of the atomic absorption spectrophotometer. Add 0.20 ml of 1-butanol (4) and mix thoroughly with an electric mini-stirrer.

Heat the extract in a graphite furnace according to an appropriate time-temperature programme (see remark). Measure the absorbance at 228.8 nm in the atomization phase, in the cleaning phase, and at room temperature at the end of the cycle; use background correction.

Plot a calibration curve and read the cadmium concentrations.

Remarks:

- Samples and standard series should be shaken at one time, or at least directly after each other, with the same shaking machine.
- When a sample gives a higher absorbance than the highest standard, it should be diluted with the extract of the zero standard solution (dilution factor = f). If a more than tenfold dilution would be required, then the extraction should be repeated with less digest pipetted (D ml).
- The calibration curve is slightly bending towards the x-axis.
- Xylene vapours are rather poisonous. Work in a fume hood.
- For samples with high Fe content, an appreciable amount of Fe can be transferred into the organic phase. This may give too low results when applying a deuterium background correction system.
- The absorbance reading in the cleaning phase is meant as a check on memory effects. If these are observed, the results of the last and the second last sample should be discarded. With the programme used in our laboratory (see remark), memory effects are only observed if something is out of order.
- After completion of each firing, the absorbance is read at room temperature as an indication of baseline drift. This value is used as a correction on the reading in the atomization phase.
- Every sample should be measured at least three times, since the precision of this technique is low.
- Clean all glassware, polythene cups and PTFE stoppers as follows: After use, wash once with ethanol 96% to remove the remaining xylene. Let drain, rinse with water and shake with 4 M HNO_3 for 5 minutes (or leave overnight). Then rinse with water and twice with ethanol 96%. Allow to dry by leaving at room temperature in an inverted position.
- The wash solution of the automatic sampler contains 2-propanol (2% in water) in order to lower its surface tension.
- The complex of Cd-DDDTC in xylene is stable for about 24 hours, and can therefore be measured on "extraction day" or the day after.
- The measurements can be performed with any ETA-AAS system. The present method was worked out using a Perkin-Elmer system, consisting of a model 460 AA spectrophotometer, a HGA 76 B graphite furnace, an AS-1 automatic injection unit annex sampler and a PRS-10 printer. With a lamp current of 6 mA, a gain of 40, an integration time of 10 s, background correction on (mode AA minus BG), slit 0.7 nm Alt and peak height mode, the optimum programme – after 30 min warming up – is as follows:

phase	I (dry)	II (char)	III (atomize)	IV (clean)	"V" (base line)
temperature (°C)	160	450	2200	2500	ambient
rate	1	0	1	0	—
hold time (s)	30	45	10	3	—
gas stop (s)	—	—	10	—	—
read	—	—	yes	yes	yes

- When using other instruments, the optimum temperature values to be set may differ from the values given above, because the temperatures mentioned here are instrument settings instead of real temperature values.

Calculation

The cadmium content of the dried plant material, expressed in µg/kg Cd, is calculated by

$$\left\{\frac{E \times a}{D} - b\right\} \times \frac{10 \times f \times V}{E \times w}$$

in which

a = concentration of cadmium in the xylene extract of the sample digest, in µg/l;

b = concentration of cadmium in the xylene extract of the blank digest, in µg/l;

f = dilution factor for the measuring solution (remark);

V = total volume of digest at the end of the digestion procedure, in ml;

D = volume of sample digest pipetted for extraction, in ml;

E = volume of blank digest pipetted for extraction, in ml; (E = highest D)

w = weight of plant material sample, in g.

Remark:

- If f = 1, D = 30, V = 100 and w = 2, then the cadmium content is (a — b)/60.

3.3.1 DETERMINATION OF **COPPER** BY FLAME AAS

Field of application

This determination is intended to be carried out on digest 3.0 (HNO_3-$HClO_4$-H_2SO_4).

Samples which are expected to be very low in Cu (< 25 mg/kg) should be determined by method 3.3.2; the present flame AAS method is especially suited for screening a series of samples in order to see which samples must be analysed by (the more cumbersome) method 3.3.2.

Principle

The sample is nebulized into an air-acetylene flame, where it is vaporized; copper compounds are atomized and the copper atoms thus formed absorb radiation from a hollow-cathode lamp. The absorbance is measured at a wavelength of 324.7 nm.

Apparatus

- Atomic absorption spectrophotometer.

Reagents

(1) Sulphuric acid, $c(H_2SO_4)$ = 0.08 mol/l. Add 4.5 ml of concentrated sulphuric acid (96%) to about 400 ml water and make up to 1 litre.

(2) Stock solution, Cu concentration 1000 mg/l. Dissolve 3.929 g of copper(II) sulphate pentahydrate, $CuSO_4 \cdot 5H_2O$ (standardized according

to Appendix A-18) in a 1000-ml volumetric flask in 0.08 M sulphuric acid (1), and make up to the mark with this acid.

(3) Standard solution, Cu concentration 50 mg/l. Pipette 25.0 ml of the stock solution (2) into a 500-ml volumetric flask and make up to volume with 0.08 M sulphuric acid (1).

Remarks:

- Frequently, Cu determinations go along with Fe determinations. In that case, a mixed standard solution may be prepared with 50 mg/l Cu and 50 mg/l Fe (see 3.4.1).

Standard series

Pipette 0–2.00–4.00–6.00–8.00–10.00 ml of the standard solution (3) into 100-ml volumetric flasks and make up to volume with 0.08 M sulphuric acid (1). This standard series has Cu concentrations of 0–1–2–3–4–5 mg/l.

Procedure

Nebulize the sample digests, the blank digests and the standard series into a just blue (stoichiometric) air-acetylene flame and measure the absorbance at a wavelength of 324.7 nm. Plot a calibration curve and read the copper concentrations.

Remarks:

- The calibration curve is slightly bent towards the x-axis, therefore the curve should not be generated by means of linear regression.
- It is not necessary to apply background correction.
- No interferences have been found.

Calculation

The copper content of the dried plant material, expressed in mg/kg Cu, is calculated by

$$(a - b) \times \frac{V}{w}$$

in which

a = concentration of copper in the sample digest, in mg/l;
b = concentration of copper in the blank digest, in mg/l;
V = total volume of digest at the end of the digestion procedure, in ml;
w = weight of plant material sample, in g.

3.3.2 DETERMINATION OF **COPPER** BY FURNACE AAS

Field of application

This determination is intended to be carried out on digest 3.0 (HNO_3-$HClO_4$-H_2SO_4).

This method is especially suited for low copper contents; since it is rather time-consuming, a preliminary screening by flame AAS (section 3.3.1) is recommended.

Principle

The sample is brought into a graphite furnace, where it is subsequently dried, ashed and vaporized by electrical heating. The copper atoms thus formed absorb radiation from a hollow-cathode lamp. The absorbance is measured at a wavelength of 324.7 nm while applying Zeeman background correction.

Remarks:

- When using Zeeman background correction, no extraction with an organic sequestering agent into xylene is needed to eliminate matrix interferences.

Apparatus

- Atomic absorption spectrophotometer, with electrothermal atomization device (graphite furnace) and with a device for correcting or measuring background absorption, preferably a Zeeman background correction system.

Reagents

(1) Sulphuric acid, $c(H_2SO_4)$ = 0.8 mol/l. Add carefully, while swirling, 45 ml of concentrated sulphuric acid (96%) to about 400 ml water. Allow to cool and make up to 1 litre.

(2) Sulphuric acid, $c(H_2SO_4)$ = 0.08 mol/l. Dilute the 0.8 M sulphuric acid (1) 1 + 9 (v/v).

(3) 1-Butanol (= n-butylalcohol).

(4) Stock solution, Cu concentration 1000 mg/l. Dissolve 3.929 g of copper(II) sulphate pentahydrate, $CuSO_4 \cdot 5H_2O$ (standardized according to App. A-18) in about 100 ml water in a 1000-ml volumetric flask. Add 100 ml of 0.8 M sulphuric acid (1) and dilute to volume.

(5) Standard solution, Cu concentration 10 mg/l. Pipette 1.00 ml of the stock solution (4) into a 100-ml volumetric flask and make up to the mark with 0.08 M sulphuric acid (2).

(6) Diluted standard solution, Cu concentration 50 μg/l. Pipette 5.00 ml of the standard solution (5) into a 1000-ml volumetric flask. Add 100 ml of 0.8 M sulphuric acid (1) and make up to volume.

Remarks:

- Frequently, Cu determinations go along with Fe determinations. If so, a mixed diluted standard solution may be prepared with 50 µg/l Cu and 50 µg/l Fe in 0.08 M sulphuric acid.

Standard series

Pipette 0–0.50–1.00–1.50–2.00 ml of the diluted standard solution (6) into 100-ml erlenmeyer flasks. Dilute to about 30 ml with 0.08 M sulphuric acid (2); mix immediately after addition of acid.

Procedure

Transfer (part of) the digests and the standard series to polythene cups that fit in the automatic sampler of the atomic absorption spectrophotometer. Add 0.20 ml of 1-butanol (3) and mix thoroughly with an electric mini-stirrer. Heat the extract in a graphite furnace according to an appropriate time-temperature programme (see remark). Measure the absorbance at 324.7 nm in the atomization phase, in the cleaning phase, and at room temperature at the end of the cycle; use Zeeman background correction. Plot a calibration curve and read the copper concentrations.

Remarks:

- When a sample gives a higher absorbance than the highest standard, it should be diluted with the zero standard solution (dilution factor = f); this can be programmed in the automatic sampler.
- The calibration curve is slightly bending towards the x-axis.
- The absorbance reading in the cleaning phase is meant as a check on memory effects. If these are observed, the results of the last and the second last sample should be discarded. With the programme used in our laboratory (see remark), memory effects are only observed if something is out of order.
- After completion of each firing, the absorbance is read at room temperature as an indication of baseline drift. This value is used as a correction on the reading in the atomization phase.
- Every sample should be measured at least three times, since the precision of this technique is low.
- The wash solution of the automatic sampler contains 2-propanol (2% in water) in order to lower its surface tension.
- The measurements can be performed with any ETA-AAS system. The present method was worked out using a Varian SpectrAA 300 system. The optimum programme – after 30 min warming up – is as follows:

phase	I (dry)	II (char)	III (atomize)	IV (clean)	"V" (base line)
temperature (°C)	160	450	2200	2500	ambient
rate	1	0	1	0	—
hold time (s)	30	45	10	3	—
gas stop (s)	—	—	10	—	—
read	—	—	yes	yes	yes

- When using other instruments, the optimum temperature values to be set may differ from the values given above, because the temperatures mentioned here are instrument settings instead of real temperature values.

Calculation

The copper content of the dried plant material, expressed in μg/kg Cu, is calculated by

$$\frac{(a - b) \times f \times V}{w}$$

in which
a = concentration of copper in the sample digest, in μg/l;
b = concentration of copper in the blank digest, in μg/l;
f = dilution factor for the measuring solution;
V = total volume of digest at the end of the digestion procedure, in ml;
w = weight of plant material sample, in g.

3.3.3 DETERMINATION OF **COPPER** BY COLORIMETRY AFTER LIQUID-LIQUID EXTRACTION

Field of application

This determination is intended to be carried out on digest 3.0 (HNO_3-$HClO_4$-H_2SO_4).

This method is especially suited for low copper contents; since it is rather time-consuming, a preliminary screening by flame AAS (section 3.3.1) is recommended.

Principle

Copper(II) ions are sequestered by a carbamate; the yellow-coloured complex is extracted from sulphuric acid medium into xylene. Its absorbance is measured at a wavelength of 435 nm.

Remarks:
- The extraction is intended to concentrate the copper, since its content in plant material is normally very low. At the same time, however, the usually interfering metals are not extracted into the organic phase because of the low pH.
- The sulphuric acid concentration should be between 0.3–1.1 mol/l.

Apparatus

- Linear shaker.
- Rack for erlenmeyer flasks to be placed in the shaker (see Fig. 6).
- Water repellent filter paper (e.g., Machery & Nagel 616 WA, or Schleicher & Schüll 597 hy, or Whatman 1PS).
- Colorimeter.

Reagents

(1) Sulphuric acid, $c(H_2SO_4) = 2.5$ mol/l. Add carefully, while swirling, 140 ml of concentrated sulphuric acid (96%) to about 400 ml water. Allow to cool and make up to 1 litre.
(2) Sulphuric acid, $c(H_2SO_4) = 0.08$ mol/l. Add 4.5 ml of concentrated sulphuric acid (96%) to about 400 ml of water and make up to 1 litre.
(3) Zinc dibenzyldithiocarbamate solution, 0.02% (w/v). Dissolve 100 mg of zinc dibenzyldithiocarbamate in 500 ml of xylene. Prepare fresh weekly.
(4) Stock solution, Cu concentration 1000 mg/l. Dissolve 3.929 g of copper(II) sulphate pentahydrate, $CuSO_4 \cdot 5H_2O$ (standardized according to App. A-18) in about 100 ml water in a 1000-ml volumetric flask. Add 32 ml of 2.5 M sulphuric acid (1) and dilute to volume.
(5) Standard solution, Cu concentration 1 mg/l. Pipette 1.00 ml of the stock solution (4) into a 1000-ml volumetric flask and make up to the mark with 0.08 M sulphuric acid (2).

Remark:
- The commonly used sodium diethyldithiocarbamate gives a less stable complex, and must be used in slightly alkaline medium. Then it forms also complexes with other metals. The zinc dibenzyldithiocarbamate, however, is stable in acid medium.

Standard series

Pipette 0–2.00–4.00–6.00–8.00–10.00 ml of the standard solution (5) into 100-ml erlenmeyer flasks that have ground glass necks and PTFE stoppers. Add respectively 20–18–16–14–12–10 ml of 0.08 M sulphuric acid (2). At this moment, this standard series has Cu concentrations of 0–0.1–0.2–0.3–0.4–0.5 mg/l.

Procedure

Extraction
Pipette D ml (D being mostly 20.00) of the sample digests and the blank digests into 100-ml erlenmeyer flasks that have ground glass necks and PTFE stoppers. Add to the standard series, the sample digests and the blank digests 10 ml of 2.5 M sulphuric acid (1). Add (40—D) ml water and mix. Then add 5.00 ml of the zinc dibenzyldithiocarbamate solution (3). Stopper the erlenmeyer flasks, place them in the rack (see Fig. 6) and shake mechanically during 10 min (stroke length 9 cm, 150 oscillations per min). Wait until the phases have separated. Suck off the water phase (see Fig. 7) to such an extent that the filter to be used can easily retain the remaining water. Filter over water repellent paper into test tubes.

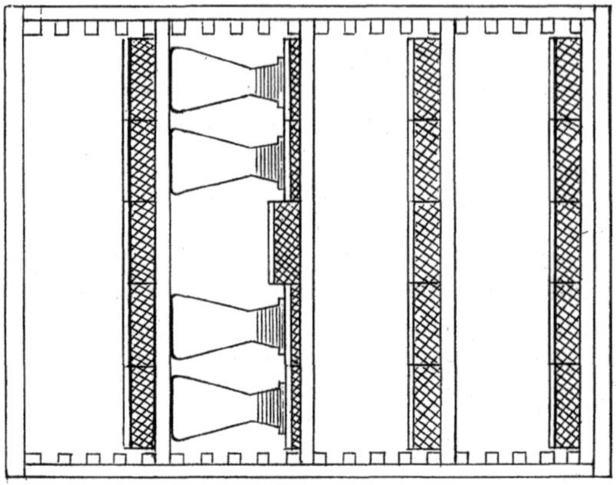

Fig. 6. Rack & erlenmeyer flasks (top view).

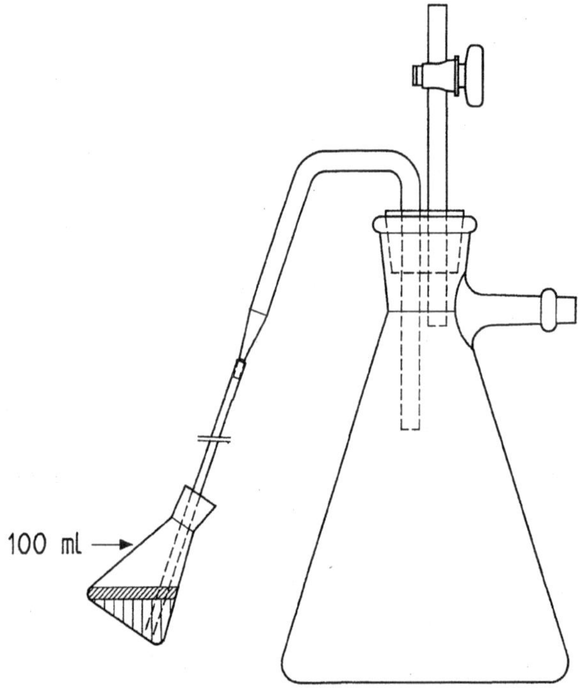

Fig. 7. Sucking off the (lower) water phase.

Measurement
Calibrate the spectrophotometer while the cuvette is filled with xylene. Measure the absorbance of all extracts in a 1-cm cuvette at a wavelength of 435 nm. Plot a calibration curve and read the copper concentrations.

Remarks:

- Samples and standard series should be shaken together at the same time, on a powerful linear shaker at the prescribed conditions. Otherwise, the extraction yield may be variable and low.
  If the conditions of 9 cm stroke length and/or 150 strokes per min cannot be fulfilled, the time of shaking should be prolonged; the optimum time should be established by shaking standard series for different periods.
- For very low concentrations, 40 ml of digest may be taken for extraction; for relatively high concentrations 5 ml will suffice.
- The colour is stable for at least 5 hours.
- The calibration curve should be linear.
- The erlenmeyer flasks should be cleaned in advance by shaking with a suspension of sodium diethyldithiocarbamate in chloroform (1% w/v). Rinse thoroughly only with demineralised water. (Chloroform and sodium diethyldithiocarbamate are used because the washing is quite easy.)
- Xylene and chloroform vapours are rather poisonous. Work in a fume hood.
- When taking 20 ml of digest 3.0, no interferences were found from concentrations of 100 mg/l Al, 5 mg/l Co, 100 mg/l Fe, 25 mg/l Mn, 10 mg/l Ni and 30 mg/l Zn in the digest. A concentration of 10 mg/l Mo causes a signal enhancement of 3% at a level of 2.5 mg/l Cu.

Calculation

The copper content of the dried plant material, expressed in mg/kg Cu, is calculated by

$$20 \times (a - b) \times \frac{1}{D} \times \frac{V}{w}$$

in which

a = concentration of copper in the xylene extract of the sample digest, in mg/l;
b = concentration of copper in the xylene extract of the blank digest, in mg/l;
D = volume of digest pipetted for extraction, in ml;
V = total volume of digest at the end of the digestion procedure, in ml;
w = weight of plant material sample, in g.

REFERENCES

S. Andrus. Determination of copper in plant materials by means of zinc dibenzyldithiocarbamate. Analyst *80* (1955) 514–516.

R.I. Martens and R.E. Githens, Sr. Small amounts of copper in dyes and rubber chemicals; colorimetric determination with zinc dibenzyldithiocarbamate. Anal. Chem. *24* (1952) 991–993.

3.4.1 DETERMINATION OF **IRON** BY FLAME AAS

Field of application

This determination is intended to be carried out on digest 3.0 (HNO_3-$HClO_4$-H_2SO_4).

The iron content of the dried plant material should be higher than 25 mg/kg.

Principle

The sample is nebulized into an air-acetylene flame, where it is vaporized; iron compounds are atomized and the iron atoms thus formed absorb radiation from a hollow-cathode lamp. The absorbance is measured at a wavelength of 248.3 nm.

Apparatus

- Atomic absorption spectrophotometer.

Reagents

(1) Sulphuric acid, $c(H_2SO_4)$ = 0.08 mol/l. Add 4.5 ml of concentrated sulphuric acid (96%) to about 400 ml water and make up to 1 litre.
(2) Stock solution, Fe concentration 1000 mg/l. Put about 200 ml water in a 1000-ml volumetric flask and add 4.5 ml of concentrated sulphuric acid (96%). Weigh out 7.022 g of iron(II) ammonium sulphate hexahydrate, $Fe(NH_4)_2(SO_4)_2 \cdot 6H_2O$, standardized according to Appendix A-19, and transfer it with some water to this 1000-ml volumetric flask. Boil to dissolve the salt completely. Allow to cool and make up to the mark.
(3) Standard solution, Fe concentration 50 mg/l. Pipette 25.0 ml of the stock solution (2) into a 500-ml volumetric flask. Make up to volume with 0.08 M sulphuric acid (1).

Remarks:

- Frequently, Fe determinations go along with Cu determinations. If so, a mixed standard solution may be prepared with 50 mg/l Fe and 50 mg/l Cu.

Standard series

Pipette 0–2.00–4.00–6.00–8.00–10.00 ml of the standard solution (3) into 100-ml volumetric flasks. Make up to volume with 0.08 M sulphuric acid (1). This standard series has Fe concentrations of 0–1–2–3–4–5 mg/l.

Procedure

Nebulize the sample digests, the blank digests and the standard series into a blue (oxidizing) air-acetylene flame and measure the absorbance at a wavelength of 248.3 nm. Plot a calibration curve and read the iron concentrations.

Remarks:

- The calibration curve is slightly bent towards the x-axis.
- It is not necessary to apply background correction.
- No interferences have been found.

Calculation

The iron content of the dried plant material, expressed in mg/kg Fe, is calculated by

$$(a - b) \times \frac{V}{w}$$

in which
a = concentration of iron in the sample digest, in mg/l;
b = concentration of iron in the blank digest, in mg/l;
V = total volume of digest at the end of the digestion procedure, in ml;
w = weight of plant material sample, in g.

3.4.2 DETERMINATION OF **IRON** BY COLORIMETRY

Field of application

This determination is intended to be carried out on digest 3.0 (HNO_3-$HClO_4$-H_2SO_4).

Principle

Iron(II) ions form an orange-red coloured complex with 1,10-phenanthroline at pH 2–9. Its absorbance is measured at a wavelength of 515 nm.

Remarks:

- Metol is used to reduce Fe(III) ions.
- When both aluminium and phosphate are present, the solution may become turbid. To prevent this, citrate is added.

Apparatus

- Colorimeter.

Reagents

(1) Sulphuric acid, $c(H_2SO_4)$ = 0.08 mol/l. Add 4.5 ml of concentrated sulphuric acid (96%) to about 400 ml water and make up to 1 litre.
(2) 1,10-Phenanthroline solution. Dissolve 0.40 g of 1,10-phenanthroline hydrochloride hydrate, $C_{12}H_8N_2HCl \cdot H_2O$, in 25 ml of ethanol 96% (v/v). Make up to 100 ml with water.
(3) Citrate solution. Dissolve 2.20 g of sodium citrate dihydrate, $Na_3C_6H_5O_7 \cdot 2H_2O$, in some water and make up to 100 ml.
(4) Metol reductant solution. Dissolve 0.10 g of metol (p-methylamino-phenol sulphate, $C_{14}H_{18}N_2O_2 \cdot H_2SO_4$), 13.7 g of sodium metabisulphite, $Na_2S_2O_5$, and 0.5 g of sodium sulphite, Na_2SO_3, in some water and make up to 100 ml.
(5) Stock solution, Fe concentration 1000 mg/l. Dissolve 7.022 g of iron(II) ammonium sulphate hexahydrate, $Fe(NH_4)_2(SO_4)_2 \cdot 6H_2O$, standardized according to App. A-19, in about 100 ml of water in a 1000-ml volumetric flask. Add 10 ml of concentrated sulphuric acid (96%). Boil to dissolve the salt completely and allow to cool down. Make up to volume with water.
(6) Standard solution, Fe concentration 100 mg/l. Pipette 50 ml of the stock solution (5) into a 500-ml volumetric flask and make up to the mark with 0.08 M sulphuric acid (1).

Standard series

Pipette 0–2.00–4.00–6.00–8.00–10.00 ml of the standard solution (6) into 100-ml volumetric flasks and make up to volume with 0.08 M sulphuric acid (1). This series has Fe concentrations of 0–2–4–6–8–10 mg/l.

Procedure

Pipette 2.00 ml of the sample digests, the blank digests and the standard series into test tubes. Add, in this sequence,

- 0.10 ml of metol reductant solution (4);
- 0.20 ml of 1,10-phenanthroline solution (2);
- 2.70 ml of citrate solution (3).

Mix after each addition. Wait for at least 1 h, then measure the absorbance in a 1-cm cuvette at a wavelength of 515 nm. Plot a calibration curve and read the Fe concentrations.

Remarks:

- The colour intensity is constant for at least 5 h.
- The calibration curve should be linear.
- Some heavy metals form complexes with 1,10-phenanthroline without showing a colour.

Therefore, it is advised to do the measurements in duplicate, and to pipette then 1.00 ml instead of 2.00 ml of the digests and to add 1.00 ml of 0.08 M sulphuric acid (1).
If the 1-ml aliquots give higher results than the 2-ml aliquots, the measurements should be repeated with 0.5-ml aliquots.
- No interferences were found from concentrations of 2000 mg/l PO_4 and 200 mg/l Al, even if present simultaneously. Furthermore, no interferences were found from 10 mg/l Co, 25 mg/l Cu, 10 mg/l Mo, 200 mg/l Ni and 100 mg/l Zn.

Calculation

The iron content of the dried plant material, expressed in mg/kg Fe, is calculated by

$$(a - b) \times \frac{V}{w}$$

in which
a = concentration of iron in the sample digest, in mg/l;
b = concentration of iron in the blank digest, in mg/l;
V = total volume of digest at the end of the digestion procedure, in ml;
w = weight of plant material sample, in g.

3.5.1 DETERMINATION OF **MANGANESE** BY FLAME AAS

Field of application

This determination is intended to be carried out on digest 3.0 (HNO_3-$HClO_4$-H_2SO_4).

The manganese content of the dried plant material should be higher than 15 mg/kg.

Principle

The sample is nebulized into an air-acetylene flame, where it is vaporized; manganese compounds are atomized and the manganese atoms thus formed absorb radiation from a hollow-cathode lamp. Lanthanum is added to prevent condensed phase interferences. The absorbance is measured at a wavelength of 279.5 nm.

Apparatus

- Atomic absorption spectrophotometer.

Reagents

(1) Sulphuric acid, $c(H_2SO_4)$ = 0.08 mol/l. Add 4.5 ml of concentrated sulphuric acid (96%) to about 400 ml water and make up to 1 litre.
(2) Lanthanum solution, La concentration 16.5 g/l. Dissolve 12.86 g of lanthanum nitrate hexahydrate, $La(NO_3)_3 \cdot 6H_2O$ in water and make up to 250 ml.
(3) Stock solution, Mn concentration 500 mg/l. Dissolve 1.438 g of potassium permanganate, $KMnO_4$, (standardized according to Appendix A-21), in about 200 ml of water in a beaker. Add carefully, while stirring, 45 ml of concentrated sulphuric acid (96%). Reduce the permanganate with a few drops of hydrogen peroxide (30%) and boil to remove the excess of H_2O_2. Transfer the contents of the beaker quantitatively to a 1000-ml volumetric flask and make up to the mark.
(4) Standard solution, Mn concentration 50 mg/l. Pipette 50.00 ml of the stock solution (3) into a 500-ml volumetric flask and make up to volume with 0.08 M sulphuric acid (1).

Remark:
- Frequently, Mn determinations go along with Zn determinations. In that case, a mixed standard solution may be prepared with 50 mg/l Mn and 20 mg/l Zn.

Standard series

Pipette 0–1.00–2.00–4.00–6.00–8.00–10.00 ml of the standard solution (4) into 100-ml volumetric flasks and make up to volume with 0.08 M sulphuric acid (1). This standard series has Mn concentrations of 0–0.5–1–2–3–4–5 mg/l.

Procedure

Pipette 5.00 ml of the sample digests, the blank digests and the standard series into test tubes. Add 0.50 ml of lanthanum solution (2) and mix. Nebulize all these solutions into a blue (oxidizing) air-acetylene flame and measure the absorbance at 279.5 nm. Plot a calibration curve and read the manganese concentrations.

Remarks:
- The calibration curve is slightly bent towards the x-axis, so that calculation by means of linear regression is not allowed.
- In the 0.08 M sulphuric acid medium, no interferences were found after spiking the digests with various salts, so as to reach final concentrations of 80 mg/l Al, 500 mg/l Ca, 80 mg/l Fe, 800 mg/l Mg or 16 mg/l Si. However, a concentration of 1000 mg/l Ca results in a precipitate of $CaSO_4$ if the solution has been left standing overnight.

Calculation

The manganese content of the dried plant material, expressed in mg/kg Mn, is calculated by

$$(a - b) \times \frac{V}{w}$$

in which

a = concentration of manganese in the sample digest, in mg/l;
b = concentration of manganese in the blank digest, in mg/l;
V = total volume of digest at the end of the digestion procedure, in ml;
w = weight of plant material sample, in g.

3.5.2 DETERMINATION OF **MANGANESE** BY COLORIMETRY

Field of application

This determination is intended to be carried out on digest 3.0 (HNO_3-$HClO_4$-H_2SO_4).

This determination is only recommended if no atomic absorption spectrophotometer is available.

Principle

In alkaline medium, manganese ions form a strongly coloured complex with formaldoxime. The absorbance of this complex is measured at a wavelength of 500 nm, after having masked any interfering metal ions by KCN.

Remark:

- Fe^{3+} is reduced by metol to Fe^{2+} and then complexed by CN^- to form the less interfering, yellow-coloured $[Fe(CN)_6]^{4-}$. This compound interferes at the wavelength of maximum absorption (450 nm), but not at 500 nm.

Apparatus

- Colorimeter with interference filter.

Reagents

(1) Sulphuric acid, $c(H_2SO_4)$ = 0.08 mol/l. Add 4.5 ml of concentrated sulphuric acid (96%) to about 400 ml water and make up to 1 litre.
(2) Metol solution. Transfer 0.10 g of p-methylaminophenolsulphate, $C_{14}H_{20}N_2O_6S$ (metol), 13.7 g of sodium metabisulphite, $Na_2S_2O_5$, and 0.5 g of sodium sulphite, Na_2SO_3, to a 100-ml volumetric flask. Add water and make up to 100 ml. Add a magnetic bar and stir until dissolved.
(3) Cyanide solution. Dissolve 2.0 g of potassium cyanide, KCN, **(CAUTION!)** in 20 ml water, and mix it with 100 ml of concentrated aqueous ammonia (25%). Then add 360 ml water.
(4) Formaldoxime solution. Add 5 g of formaldehyde solution (37%) to 80 ml water. Dissolve 5 g of hydroxylamine-HCL in this mixture. Dilute to 100 ml.

(5) Standard solution, Mn concentration 100 mg/l. Dissolve 0.2877 g of potassium permanganate, $KMnO_4$, (standardized according to App. A-21) in about 200 ml water in a beaker. Add 4.5 ml of concentrated sulphuric acid (96%) and mix. Reduce the permanganate with a few drops of hydrogen peroxide (30%) and boil to remove the excess of H_2O_2. Allow to cool; transfer the contents of the beaker quantitatively to a 1000-ml volumetric flask and make up to the mark.

Standard series

Pipette 0–1.00–2.00–4.00–6.00–8.00–10.00 ml of the standard solution (5) into 100-ml volumetric flasks and make up to volume with 0.08 M sulphuric acid (1). This standard series has Mn concentrations of 0–1–2–4–6–8–10 mg/l.

Procedure

Pipette 2.00 ml of the sample digests, the blank digests and the standard series into test tubes. For duplicate measurements, pipette 1.00 ml of the digests (*not* the standards) into test tubes and add 1.00 ml of 0.08 M sulphuric acid (1).
Place the racks with the test tubes in the fume cupboard.
Add to all test tubes, in this sequence:

- 0.50 ml metol solution (2) and mix;
- 2.40 ml cyanide solution (3); mix and wait for 15 min;
- 2 drops of formaldoxime solution (4); mix immediately and wait for 30 min.

Measure the absorbance at a wavelength of 500 nm in a 1-cm cuvette. Plot a calibration curve and read the manganese concentrations.

Remarks:

- The cyanide solution is very poisonous. When it is added to the (acidic) digests, HCN may be released, therefore the addition should be done in a fume cupboard. The measurements should be done preferably in the fume cupboard also. After measuring, all CN containing solutions should be collected for appropriate processing.
- If turbidity occurs in the sample digests after the reagents have been added, a knife point of Na_2EDTA may be added. If the turbidity persists, 0.5 ml of sample should be pipetted.
- After the development period, the colour is constant for at least 5 hours.
- The calibration curve should be linear.
- In the 0.08 M sulphuric acid medium no interferences have been found from 50 mg/l Co, 50 mg/l Cu, 50 mg/l Ni or 50 mg/l Zn.

Calculation

The manganese content of the dried plant material, expressed in mg/kg Mn, is calculated by

$$(a - b) \times f \times \frac{V}{w}$$

in which

a = concentration of manganese in the sample digest, in mg/l;
b = concentration of manganese in the blank digest, in mg/l;
f = dilution factor: f = 2 when 1 ml was pipetted; f = 1 when 2 ml were pipetted;
V = total volume of digest at the end of the digestion procedure, in ml;
w = weight of plant material sample, in g.

REFERENCES

J.Ch. van Schouwenburg. The routine determination of manganese with formaldoxime. Neth. J. agric. Sci. *14* (1966) 127–130.

3.6.1 DETERMINATION OF **LEAD** BY FLAME AAS

Field of application

This determination is intended to be carried out on digest 3.0 (HNO_3-$HClO_4$-H_2SO_4).

The lead content of the dried plant material should be higher than 25 mg/kg. Samples with lower Pb levels should be determined by method 3.6.2; since this method is more cumbersome, it is recommended to apply method 3.6.1 first to all samples as a screening.

Principle

The sample is nebulized into an air-acetylene flame, where it is vaporized; lead compounds are atomized and the lead atoms thus formed absorb radiation from a hollow-cathode lamp. The absorbance is measured at a wavelength of 217.0 nm.

Remarks:

- The plant material digest appears to create a continuous background absorption, so that background correction is necessary.
- Extremely high Fe levels (> 5 g/kg dry material) will cause noticeable spectral interference because of a structured background. When using deuterium background correction, this may lead to too low analytical results (see literature reference).

Apparatus

- Atomic absorption spectrophotometer, with a device for correcting or measuring background absorption.

Reagents

(1) Sulphuric acid, $c(H_2SO_4) = 0.8$ mol/l. Add carefully, while swirling, 45 ml of concentrated sulphuric acid (96%) to about 400 ml water. Allow to cool and make up to 1 litre.

(2) Stock solution, Pb concentration 1000 mg/l. Dissolve 1.5985 g of lead nitrate, $Pb(NO_3)_2$ (standardized according to Appendix A-23) in some water in a 1000-ml volumetric flask. Add 10 ml of concentrated nitric acid (65%) and make up to the mark.

(3) Standard solution, Pb concentration 100 mg/l. Pipette 10.0 ml of the stock solution (2) into a 100-ml volumetric flask and make up to the mark.

Remarks:
- Neither the stock solution (2) nor the standard solution (3) should be made in sulphuric acid, because a precipitate of PbSO4 would be formed. Only when preparing the standard series, the acid type of the digest should be considered.
- Frequently, Pb determinations go along with Cd determinations. In that case, a mixed standard solution may be prepared with 100 mg/l Pb and 10 mg/l Cd (see remark above).

Standard series

Pipette 0–0.50–1.00–2.00–3.00–4.00 ml of the standard solution (3) into 100-ml volumetric flasks. Add about 70 ml water and 10 ml of 0.8 M sulphuric acid (1). Make up to volume with water. This standard series has Pb concentrations of 0–0.5–1–2–3–4 mg/l.

Remark:
- By first adding the bulk of water, one makes sure that the solubility product of $PbSO_4$ is not surpassed in the sulphuric acid standards. The amount of nitric acid may be neglected.

Procedure

Nebulize the sample digests, the blank digests and the standard series into a just blue (stoichiometric) air-acetylene flame and measure the absorbance at 217.0 nm. Plot a calibration curve and read the lead concentrations.

Remarks:
- The calibration curve is slightly bent towards the x-axis, therefore the curve should not be generated by means of linear regression.
- Background correction should be applied.
- Lead absorbance can also be measured at a wavelength of 283.3 nm. In that case, the sensitivity is only half of that at 217.0 nm, whereas the noise and the background do not differ appreciably.

Calculation

The lead content of the dried plant material, expressed in mg/kg Pb, is calculated by

$$(a - b) \times \frac{V}{w}$$

in which
a = concentration of lead in the sample digest, in mg/l;
b = concentration of lead in the blank digest, in mg/l;
V = total volume of digest at the end of the digestion procedure, in ml;
w = weight of plant material sample, in g.

REFERENCES

J.J. van der Lee, E. Temminghoff, V.J.G. Houba and I. Novozamsky. Background corrections in the determination of Cd and Pb by flame AAS in plant and soil samples with high Fe levels. Appl. Spectrosc. *41* (1987) 388–390.

3.6.2 DETERMINATION OF LEAD BY FURNACE AAS

Field of application

This determination is intended to be carried out on digest 3.0 (HNO_3-$HClO_4$-H_2SO_4).

It is intended for samples that are very low in Pb (< 25 mg/kg).

Principle

The sample is brought into a graphite furnace, where it is subsequently dried, ashed and vaporized by electrical heating. The lead atoms thus formed absorb radiation from a hollow-cathode lamp. The absorbance is measured at a wavelength of 283.3 nm while applying Zeeman background correction.

Remarks:
- When using Zeeman background correction, no extraction with an organic sequestering agent into xylene is needed to eliminate matrix interferences.
- Whereas the flame AAS determination (3.6.1) uses 217.0 nm as wavelength of measurement, the graphite furnace determination has a better signal-to-noise ratio at 283.3 nm and less background.

Apparatus

- Atomic absorption spectrophotometer, with electrothermal atomization device (graphite furnace) and with a device for correcting or measuring background absorption, preferably a Zeeman background correction system.

Reagents

(1) Sulphuric acid, $c(H_2SO_4)$ = 0.08 mol/l. Add 4.5 ml of concentrated sulphuric acid (96%) to about 400 ml water and make up to 1 litre.
(2) 1-Butanol (= n-butylalcohol).
(3) Stock solution, Pb concentration 1000 mg/l. Dissolve 1.5985 g of lead nitrate, $Pb(NO_3)_2$, standardized according to Appendix A-23, in some water in a 1000-ml volumetric flask. Add 10 ml of concentrated nitric acid (65%) and make up to the mark.
(4) Standard solution, Pb concentration 1 mg/l. Pipette 1.00 ml of the stock solution (3) into a 1000-ml volumetric flask. Add 1 ml of concentrated nitric acid (65%) and make up to the mark with water.

Remark:
- Frequently, Pb determinations go along with Cd determinations. If so, a mixed diluted standard solution may be prepared with 1000 µg/l Pb and 50 µg/l Cd in 0.08 M sulphuric acid.

Standard series

Pipette 0–0.50–1.00–1.50–2.00 ml of the standard solution (4) into 100-ml erlenmeyer flasks. If Cd is also to be determined, pipette 0–0.50–1.00–1.50–2.00 ml of a 50 µg/l Cd solution (see 3.2.2) into the same erlenmeyer flasks. Dilute to about 30 ml with 0.08 M sulphuric acid (1); mix immediately after addition of acid.

Procedure

Transfer (part of) the digests and the standard series to polythene cups that fit in the automatic sampler of the atomic absorption spectrophotometer. Add 0.20 ml of 1-butanol (3) and mix thoroughly with an electric mini-stirrer. Heat the extract in a graphite furnace according to an appropriate time-temperature programme (see remark). Measure the absorbance at 283.3 nm in the atomization phase, in the cleaning phase, and at room temperature at the end of the cycle; use Zeeman background correction. Plot a calibration curve and read the lead concentrations.

Remarks:
- When a sample gives a higher absorbance than the highest standard, it should be diluted with the zero standard solution (dilution factor = f); this can be programmed in the automatic sampler.
- The calibration curve is slightly bending towards the x-axis.
- Samples with high Fe content may give too low results when applying a deuterium background correction system.
- The absorbance reading in the cleaning phase is meant as a check on memory effects. If these are observed, the results of the last and the second last sample should be discarded. With the programme used in our laboratory (see remark), memory effects are only observed if something is out of order.

- After completion of each firing, the absorbance is read at room temperature as an indication of baseline drift. This value is used as a correction on the reading in the atomization phase.
- Every sample should be measured at least three times, since the precision of this technique is low.
- The wash solution of the automatic sampler contains 2-propanol (2% in water) in order to lower its surface tension.
- The measurements can be performed with any ETA-AAS system. The present method was worked out using a Varian SpectrAA 300 system. The optimum programme – after 30 min warming up – is as follows:

phase	I (dry)	II (char)	III (atomize)	IV (clean)	"V" (base line)
temperature (°C)	160	500	2300	2500	ambient
rate	1	0	0	0	—
hold time (s)	30	45	10	3	—
gas stop (s)	—	—	10	—	—
read	—	—	yes	yes	yes

- When using other instruments, the optimum temperature values to be set may differ from the values given above, because the temperatures mentioned here are instrument settings instead of real temperature values.

Calculation

The lead content of the dried plant material, expressed in mg/kg Pb, is calculated by

$$\frac{(a - b) \times f \times V}{w}$$

in which
a = concentration of lead in the sample digest, in μg/l;
b = concentration of lead in the blank digest, in μg/l;
f = dilution factor for the measuring solution;
V = total volume of digest at the end of the digestion procedure, in ml;
w = weight of plant material sample, in g.

3.6.3 DETERMINATION OF **LEAD** BY FURNACE AAS AFTER LIQUID-LIQUID EXTRACTION

Field of application

This determination is intended to be carried out on digest 3.0 (HNO_3-$HClO_4$-H_2SO_4).

It is intended for samples that are very low in Pb (< 25 mg/kg).
Method 3.6.3 should only be applied when no Zeeman background correction system is available.

Principle

Lead ions are sequestered by a carbamate, and the sample is extracted into xylene. The extract is subsequently dried, ashed and vaporized by electrical heating in a graphite furnace. The lead atoms thus formed absorb radiation from a hollow-cathode lamp. The absorbance is measured at a wavelength of 283.3 nm.

Remarks:

- The extraction serves two purposes at a time: a large part of the matrix interferences is eliminated, and the lead is more concentrated.
- Since not all interferences will have disappeared, background correction is necessary.
- Whereas the flame AAS determination (3.6.1) uses 217.0 nm as wavelength of measurement, the graphite furnace determination has a better signal-to-noise ratio at 283.3 nm and less background.

Apparatus

- Erlenmeyer flasks with ground glass necks and PTFE stoppers.
- Linear shaker.
- Water repellent filter paper.
- Polythene cups.
- Atomic absorption spectrophotometer, with electrothermal atomization device (graphite furnace) and with a device for correcting or measuring background absorption.

Reagents

(1) Sulphuric acid, $c(H_2SO_4)$ = 0.08 mol/l. Add 4.5 ml of concentrated sulphuric acid (96%) to about 400 ml water and make up to 1 litre.
(2) DDDTC solution. Dissolve 5 g of diethylammonium-N,N-diethyldithiocarbamate, $C_9H_{22}N_2S_2$, in 100 ml xylene. Stir with a magnetic stirrer until all crystals are dissolved; this will take about 45 minutes. The solution can be kept overnight at room temperature.
(3) 1-Butanol (= n-butylalcohol).
(4) Stock solution, Pb concentration 1000 mg/l. Dissolve 1.5985 g of lead nitrate, $Pb(NO_3)_2$, standardized according to Appendix A-23, in some water in a 1000-ml volumetric flask. Add 10 ml of concentrated nitric acid (65%) and make up to the mark.
(5) Standard solution, Pb concentration 1 mg/l. Pipette 1.00 ml of the stock solution (4) into a 1000-ml volumetric flask. Add 1 ml of concentrated nitric acid (65%) and make up to the mark with water.

Remarks:

- To prevent precipitation of $PbSO_4$, the stock solution (4) and the standard solution (5) are acidified with HNO_3. Only the standard series will be made with the acid to match the digest.
- Frequently, Pb determinations go along with Cd determinations. In that case, a mixed standard series may be prepared as described under Standard series.
- The solid DDDTC should be stored in a refrigerator in order to prevent deterioration.

Standard series

Pipette 0–0.50–1.00–1.50–2.00 ml of the standard solution (5) into 100-ml erlenmeyer flasks. If Cd is also to be determined, pipette 0–0.50–1.00–1.50–2.00 ml of a 50 μg/l Cd solution into the same erlenmeyer flasks. Dilute to about 30 ml with 0.08 M sulphuric acid (1); mix immediately after addition of acid.

Procedure

(a) Pipetting of digests
Pipette D ml (D varying between 1 and 30 ml, according to the expected lead concentrations) of the sample digests into 100-ml erlenmeyer flasks; pipette also E ml (E = highest D) of the blank digests into 100-ml erlenmeyer flasks. Dilute to 30 ml with 0.08 M sulphuric acid (1).

(b) Extraction
Add both to standard series, sample digests and blank digests 10.00 ml of the DDDTC solution (2) and close the erlenmeyer flasks with a PTFE stopper. Shake mechanically (see Fig. 8) during 4 min (stroke length 9 cm, 150 oscillations/min). Suck off the water phase (see Fig. 9) to such an extent that the remaining water will not occupy more than half of the filter area, and filter over water repellent filter paper.

Shake an extra amount of five portions of the zero standard for future dilution purposes; mix the five extracts to make a uniform zero standard (pool).

After the extraction, the standard series has lead concentrations of 0–50–100–150–200 μg/l. (If cadmium was also put in, its concentrations are 0–2.5–5–7.5–10 μg/l).

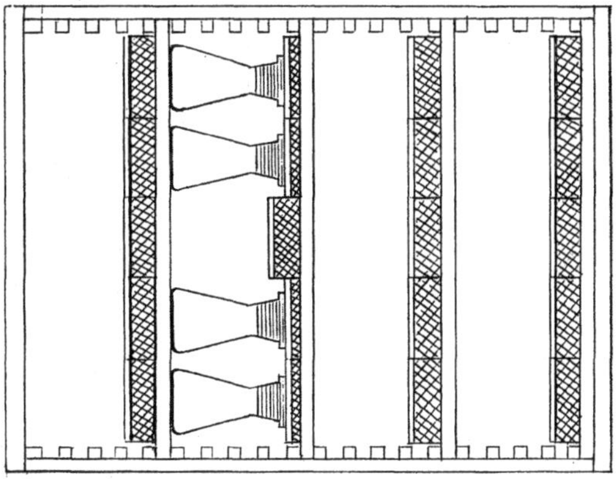

Fig. 8. Rack & erlenmeyer flasks (top view).

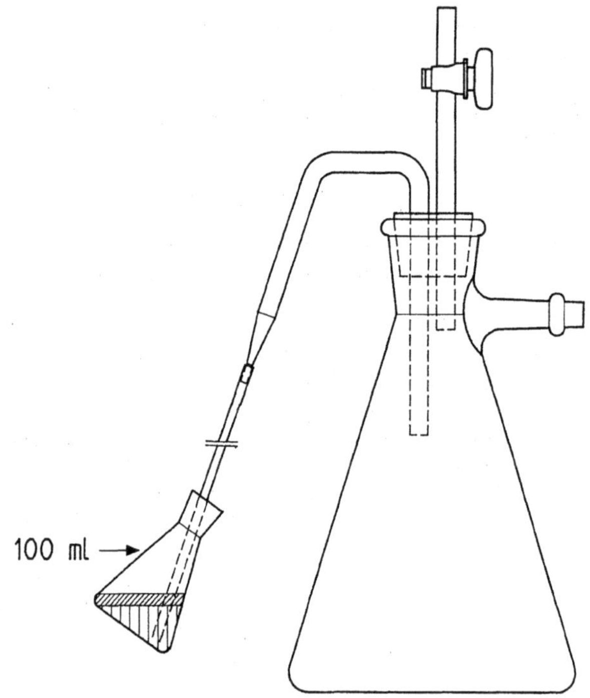

Fig. 9. Sucking off the (lower) water phase.

(c) Measurement

Pipette 1.00 ml of the extracts obtained at (b) into polythene cups that fit in the automatic sampler of the atomic absorption spectrophotometer. Add 0.20 ml of 1-butanol (3) and mix thoroughly with an electric mini-stirrer.

Heat the extract in a graphite furnace according to an appropriate time-temperature programme (see remark). Measure the absorbance at 283.3 nm in the atomization phase, in the cleaning phase, and at room temperature at the end of the cycle; use background correction.

Plot a calibration curve and read the lead concentrations.

Remarks:

- Samples and standard series should be shaken at one time, or at least directly after each other, with the same shaking machine.
- When a sample gives a higher absorbance than the highest standard, it should be diluted with the extract of the zero standard solution (dilution factor = f). If a more than tenfold dilution would be required, then the extraction should be repeated with less digest pipetted (D ml).
- The calibration curve is slightly bending towards the x-axis.
- Xylene vapours are rather poisonous. Work in a fume hood.
- For samples with high Fe content, an appreciable amount of Fe can be transferred into the organic phase. This may give too low results when applying a deuterium background correction system.
- The absorbance reading in the cleaning phase is meant as a check on memory effects. If these are observed, the results of the last and the second last sample should be discarded.

With the programme used in our laboratory (see remark), memory effects are only observed if something is out of order.

- After completion of each firing, the absorbance is read at room temperature as an indication of baseline drift. This value is used as a correction on the reading in the atomization phase.
- Every sample should be measured at least three times, since the precision of this technique is low.
- Clean all glassware, polythene cups and PTFE stoppers as follows: After use, wash once with ethanol 96% to remove the remaining xylene. Let drain, rinse with water and shake with 4 M HNO_3 for 5 minutes (or leave overnight). Then rinse with water and twice with ethanol 96%. Allow to dry by leaving at room temperature in an inverted position.
- The wash solution of the automatic sampler contains 2-propanol (2% in water) in order to lower its surface tension.
- The complex of Pb-DDDTC in xylene is stable for about 24 hours, and can therefore be measured on "extraction day" or the day after.
- The measurements can be performed with any ETA-AAS system. The present method was worked out using a Perkin-Elmer system, consisting of a model 460 AA spectrophotometer, a HGA 76 B graphite furnace, an AS-1 automatic injection unit annex sampler and a PRS-10 printer. With a lamp current of 5 mA, a gain of 40, an integration time of 10 s, background correction on (mode AA minus BG), slit 0.7 nm Alt and peak height mode, the optimum programme – after 30 min warming up – is as follows:

phase	I (dry)	II (char)	III (atomize)	IV (clean)	"V" (base line)
temperature (°C)	160	500	2300	2500	ambient
rate	1	0	0	0	—
hold time (s)	30	45	10	3	—
gas stop (s)	—	—	10	—	—
read	—	—	yes	yes	yes

- When using other instruments, the optimum temperature values to be set may differ from the values given above, because the temperatures mentioned here are instrument settings instead of real temperature values.

Calculation

The lead content of the dried plant material, expressed in µg/kg Pb, is calculated by

$$\left\{\frac{E \times a}{D} - b\right\} \times \frac{10 \times f \times V}{E \times w}$$

in which

a = concentration of lead in the xylene extract of the sample digest, in µg/l;
b = concentration of lead in the xylene extract of the blank digest, in µg/l;
f = dilution factor for the measuring solution (remark);
V = total volume of digest at the end of the digestion procedure, in ml;
D = volume of sample digest pipetted for extraction, in ml;
E = volume of blank digest pipetted for extraction, in ml; (E = highest D)
w = weight of plant material sample, in g.

Remark:

- If f = 1, D = 30, V = 100 and w = 2, then the lead content is (a — b)/60.

3.7 DETERMINATION OF **ZINC** BY FLAME AAS

Field of application

This determination is intended to be carried out on digest 3.0 (HNO_3-$HClO_4$-H_2SO_4).

The zinc content of the dried plant material should be higher than 4.5 mg/kg.

Principle

The sample is nebulized into an air-acetylene flame, where it is vaporized; zinc compounds are atomized and the zinc atoms thus formed absorb radiation from a hollow-cathode lamp. The absorbance is measured at a wavelength of 213.9 nm.

Apparatus

- Atomic absorption spectrophotometer.

Reagents

(1) Sulphuric acid, $c(H_2SO_4)$ = 0.08 mol/l. Add 4.5 ml of concentrated sulphuric acid (96%) to about 400 ml water and make up to 1 litre.
(2) Stock solution, Zn concentration 400 mg/l. Dissolve 1.759 g of zinc sulphate heptahydrate, $ZnSO_4 \cdot 7H_2O$ (standardized according to Appendix A-24) in about 600 ml water in a 1000-ml volumetric flask. Add 45 ml of concentrated sulphuric acid (96%), allow to cool and make up to volume.
(3) Standard solution, Zn concentration 20 mg/l. Pipette 25.0 ml of the stock solution (2) into a 500-ml volumetric flask and make up to volume with 0.08 M sulphuric acid (1).

Remark:
- Frequently, Zn determinations go along with Mn determinations. In that case, a mixed standard solution may be prepared with 20 mg/l Zn and 50 mg/l Mn.

Standard series

Pipette 0–1.00–2.00–4.00–6.00–8.00–10.00 ml of the standard solution (3) into 100-ml volumetric flasks and make up to volume with 0.08 M sulphuric acid (1). This standard series has Zn concentrations of 0–0.2–0.4–0.8–1.2–1.6–2.0 mg/l.

Procedure

Nebulize the sample digests, the blank digests and the standard series into a just blue (stoichiometric) air-acetylene flame and measure the absorbance at 213.9 nm. Plot a calibration curve and read the zinc concentrations.

Remarks:

- The calibration curve is slightly bent towards the x-axis, so that calculation by means of linear regression is not allowed.
- Although the chosen wavelength is short, no background correction has been found necessary when using 0.08 M sulphuric acid, 0.25 M sulphuric acid or 0.15 M perchloric acid. For other media, background correction may be required.
- No interferences were observed after spiking the digests with several salts, so as to reach final concentrations of 100 mg/l Mg, 30 mg/l Fe, 20 mg/l Al + 400 mg/l PO_4, or 30 mg/l Fe + 20 mg/l Al + 400 mg/l PO_4.
- Do not use rubber stoppers, since these will release zinc.

Calculation

The zinc content of the dried plant material, expressed in mg/kg Zn, is calculated by

$$(a - b) \times \frac{V}{w}$$

in which

a = concentration of zinc in the sample digest, in mg/l;
b = concentration of zinc in the blank digest, in mg/l;
V = total volume of digest at the end of the digestion procedure, in ml;
w = weight of plant material sample, in g.

4 Digestion with HNO_3 and determination of S

4.0 Digestion with HNO_3
4.1 Determination of **sulphur** by ICP-ES

4.0 DIGESTION WITH HNO_3

Field of application

This digestion was especially developed for the ICP determination of S-total.

Principle

Boiling nitric acid oxidizes most of the organic matter.

Remark:
- This digestion is incomplete, but the extremely high temperature of the ICP causes all undigested compounds to atomize.

Apparatus

- Metal weighing funnel (see Fig.1).
- Conventional hot plate.

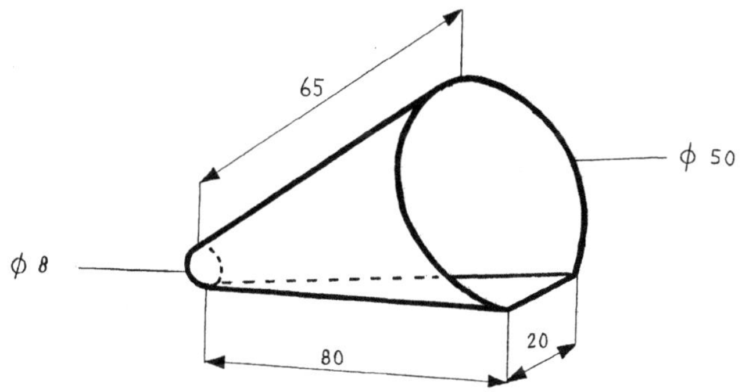

Fig. 1. Weighing funnel.

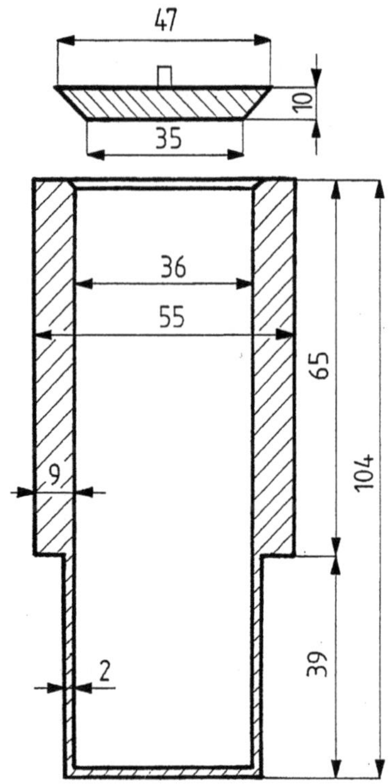

measures in mm

Fig. 2. Digestion pot.

- Aluminium heating block, 50 mm thick, with cylindrical holes of 41 mm diameter and 39 mm deep, to be laid on the hot plate.
- PTFE pots, home-made, precisely fitting the heating block holes (see Fig. 2).

Remarks:

- Other heating blocks and pots (or tubes) may be used, provided that the pots/tubes just fit the holes for efficient heat transfer, and rise high enough above the block for cooling.
- Dried plant material may easily stick to glass when the relative humidity of the air is low. These home-made weighing funnels (stainless steel or aluminium) do not show this effect.

Reagents

(1) Nitric acid, 65% (w/w), c(HNO_3) = 14.4 mol/l (d = 1.40 g/ml).
(2) Nitric acid, c(HNO_3) = 4 mol/l. Add 277 ml of nitric acid (1) to about 400 ml water and make up to 1 litre.

Procedure

Weigh, to the nearest 0.001 g, approximately 1 g of the dried plant material sample in a metal weighing funnel and transfer the sample to the PTFE pot provided. Add 10 ml of concentrated nitric acid (1) and swirl the pot carefully to moisten all the plant material. Cover the pot with its lid and allow to stand overnight. Prepare also a blank digestion.

Place the pot in a hole of the heating block and boil for 4 h at 120 °C. Remove the lid and allow the liquid to evaporate until about 1 ml remains. Take up the residue in 5 ml of nitric acid (2), boil for 10 min, let cool, transfer to a 100-ml volumetric flask and make up to the mark. Filter over fine paper.

REFERENCES

I. Novozamsky, R. van Eck, J.J. van der Lee, V.J.G. Houba and E. Temminghoff. Determination of total sulphur and extractable sulphate in plant materials by inductively-coupled plasma atomic emission spectrometry. Commun. Soil Sci. Plant Anal. 17 (1986) 1147–1157.

4.1 DETERMINATION OF **SULPHUR** BY ICP-ES

Field of application

This determination is intended to be carried out on digest 4.0 (HNO_3).

The sulphur content of the dried plant material should be higher than 8 mmol/kg.

Principle

Solutions with sulphur compounds are nebulized into an argon plasma, where all components are vaporized. Sulphur compounds are dissociated and excited, and then emit radiation of which the intensity is measured at a wavelength of 182.04 nm.

Remark:

- The short wavelength requires a monochromator that can be evacuated, or flushed with nitrogen gas, to remove interfering oxygen.

Apparatus

- Inductively-coupled plasma emission spectrometer, with a monochromator that can either be evacuated or flushed with an inert gas.

Reagents

(1) Nitric acid, $c(HNO_3)$ = 4 mol/l. Add 277 ml of concentrated nitric acid (65%) to about 600 ml of water and make up to 1 litre.

(2) Nitric acid, $c(HNO_3)$ = 0.2 mol/l. Dilute 50 ml of 4 M nitric acid (1) to 1 litre.
(3) Stock solution, S concentration 1250 mg/l. Dissolve 6.793 g of potassium sulphate, K_2SO_4, in some water in a 1000-ml volumetric flask. Add 50 ml of 4 M nitric acid (1) and make up to volume.

Standard series

Pipette 0–4.00–8.00–12.00 ml of the stock solution (3) into 100-ml volumetric flasks. Make up to the mark with 0.2 M nitric acid (2). This standard series has S concentrations of 0–50–100–150 mg/l.

Procedure

(a) Calibration of the spectrometer
Flush the monochromator with nitrogen gas for 2 hours. Then ignite the plasma and continue the flushing until the measurements are finished. Let the plasma stabilize for at least one hour before the measurements will be made.

Set the wavelength of the spectrometer at 182.04 nm; fix the other operating parameters (power, gas flows, observation height, integration time, read delay etc.) at their – previously elaborated – appropriate values for sulphur determination.

Nebulize, in this order, the highest standard (150 mg/l) and the zero standard, and enter their nominal concentration values. The spectrometer is now calibrated in concentration units.

Next, measure the concentrations of the highest standard and the zero standard as if it were samples, to check the calibration procedure. When the values thus found deviate more than 10% from the nominal values, the calibration procedure should be repeated. In the other case, measure the concentration of the rest of the standard series to check the required linearity.

(b) Measurement
Nebulize the sample digests and the blank digests and measure the sulphur concentrations immediately after calibration. With intervals of 9 samples, measure the highest standard and the zero standard again. Correct the results for the samples arithmetically for any zero drift or change in sensitivity.

Remarks:
- In the course of the day, the sensitivity changes continuously. With the authors' apparatus this change is seldom larger than 10% over a 9-sample interval. In this case, the arithmetical correction procedure is better than recalibration.
- No interference was observed when nebulizing 1000 mg/l-solutions of Al, As, Ba, Ca, Cd, Cr, Cu, Fe, K, Mg, Mn, Na, Ni, P, Sr and Ti, and 100 mg/l-solutions of B and Mo.
- In the authors' laboratory, the limit of determination (for 5% C.V., based on quadruplate instrument readings) proved to be 2.7 mg/l, corresponding with a sulphur content of 8.4 mmol/kg dry plant material. This is sufficiently low for almost any plant material.

Calculation

The sulphur content of the dried plant material, expressed in mmol/kg S, is calculated by

$$0.0312 \times (a - b) \times \frac{V}{w}$$

in which
a = concentration of sulphur in the sample digest, in mg/l;
b = concentration of sulphur in the blank digest, in mg/l;
V = total volume of digest at the end of the digestion procedure, in ml;
w = weight of plant material sample, in g.

REFERENCES

I. Novozamsky, R. van Eck, J.J. van der Lee, V.J.G. Houba and E. Temminghoff. Determination of total sulphur and extractable sulphate in plant materials by inductively-coupled plasma atomic emission spectrometry. Commun. Soil Sci. Plant Anal. *17* (1986) 1147–1157.

5 Digestion by **dry-ashing** followed by treatment with HF and determination of Ca, Cd, Cu, Fe, K, Mn, Na, Pb, Zn

5.0 DIGESTION BY **DRY-ASHING** FOLLOWED BY TREATMENT WITH HF

Field of application

With this dry-ashing method a digest is obtained in which many elements can be determined, except B, N and Si. The method is especially suitable for determination of trace elements, since large amounts of plant material can be worked up.

Principle

The organic matrix of the plant material is destroyed by controlled heating. Silicates are removed as volatile fluorides.

Remarks:
- The required high temperature causes various nitrogen compounds to volatilize.
- Treatment with HF is necessary to release any compounds adsorbed to or included in silicates. As a consequence, Si and B cannot be determined.

Apparatus

- Platinum or sintered aluminium oxide ("Alsint") crucibles.
- Muffle furnace.
- Hot plate.
- Stirring rods.

Reagents

(1) Hydrofluoric acid, 40% (w/w) , c(HF) = 22.6 mol/l (d = 1.13 g/ml).
(2) Hydrochloric acid, 37% (w/w) , c(HCl) = 12 mol/l (d = 1.19 g/ml).

Remark:
- Hydrofluoric acid is a treacherous skin poison. Work in a fume hood and wear rubber gloves, safety goggles and protective cloths. If any HF comes into contact with the skin, wash immediately and thoroughly with water and thereafter dab with calcium gluconate gel.

Procedure

Weigh, to the nearest 0.001 g, approximately 2 g of the dried plant material sample in a platinum or Alsint crucible. Place it in a muffle furnace at room temperature. Prepare also a blank digestion.

Raise the temperature gradually in 2 h to 450 °C, and maintain this temperature during 2 h. Take the crucible out of the furnace and allow it to cool e.g. on top of a tripod stand covered with Ni-Cr wire gauze.

Add 2–3 ml of water and then carefully 1 ml of hydrochloric acid (2). Wait until the production of gas bubbles has stopped. Heat gently for some minutes on a hot plate (max. 80 °C) until the appearance of fumes. Cool, add 2–3 ml of water and filter over fine paper into a 100-ml volumetric flask. Wash with luke-warm water till acid-free (< 50 ml total).

Put the filter paper and its contents into the same platinum or Alsint crucible and warm it gently in an oven until dry. Then place the crucible in the cold muffle furnace and raise the temperature to 550 °C. Maintain this temperature during 0.5 h. Take the crucible out of the furnace and allow to cool.

Add 5 ml of hydrofluoric acid (1). Heat the crucible on a hot plate until dry, taking care that the liquid does not boil (temperature about 100 °C). Allow to cool. Add 1 ml of hydrochloric acid (2), heat until appearance of fumes, cool, add 2–3 ml of water and filter into the same volumetric flask. Wash until acid-free. Make up to the mark.

Remarks:

- See Appendix A-7 for correct use and treatment of platinum crucibles.
- Platinum crucibles should be cleaned as follows: Add to the crucibles a spoonful of potassium hydrogen sulphate. Heat on top of a blue flame until the salt has dissolved. Rotate the crucible while heating, making use of a crucible tong with platinum lips. Allow to cool and dissolve the salt in warm water. Rinse with water to remove the salt completely. *Note*: Alsint crucibles should be cleaned with hot nitric acid.
- With Alsint crucibles the HCl procedure should be finished rapidly since the liquid "creeps" over the rim of the crucible.

REFERENCES

Membres du C.I.I. Méthodes de référence pour la détermination des éléments minéraux dans les végétaux. L'Agronomie Tropicale *24* (1969) 827–835. (in French)

5.1 DETERMINATION OF **CALCIUM** BY COMPLEXOMETRIC TITRATION

Field of application

This determination is intended to be carried out on digest 5.0 (dry-ashing followed by HF treatment).

This titration is only recommended in case no flame AES or flame AAS is available.

Principle

Calcium ions are titrated with EDTA using HHSNN as a combined metal ion/acid-base indicator.

Remarks:

- Interfering ions are masked by cyanide and by triethanolamine.
- With magnesium ions added, a much clearer end point is obtained; the magnesium hydroxide formed is kept in colloidal state by a colloid protector.
- HHSNN is a very sensitive indicator, which – in the presence of Ca and Mg – is coloured red-violet from pH 4–8.5, blue from pH 9–9.5, red-violet from pH 10.5–11.5, blue-violet at pH 11.7 and red at pH 12.1 and higher.

Apparatus

- Magnetic stirrer.
- Plunger-type burette, or 25-ml burette.

Reagents

(1) Hydrochloric acid, c(HCl) = 1.0 mol/l. Add 8.3 ml of concentrated hydrochloric acid (36%) to about 40 ml water and make up to 100 ml.
(2) Potassium hydroxide solution, c(KOH) = 2 mol/l. Dissolve 56.1 g of potassium hydroxide, KOH, in about 400 ml of water. Allow to cool and make up to 500 ml.
(3) Magnesium additive. Dissolve 0.384 g of magnesium sulphate heptahydrate, $MgSO_4 \cdot 7H_2O$, in about 450 ml of water in a beaker. Boil the solution and add, while stirring, 12.5 g of sodium carboxymethylcellulose (for example, Nymcel ZLF 20; see remark). Remove the beaker from the source of heat and continue stirring until the larger particles have dissolved and only a slightly turbid solution is remaining. Allow to cool further and make up to 500 ml. Filter on a Büchner funnel over medium filter paper. When the filtration speed ceases too much, a new filter paper should be installed.
(4) Masking solution. Transfer 31 ml of triethanolamine, $C_6H_{15}NO_3$, to a 250-ml volumetric flask. Because of its high viscosity, this should be done by means of 3 portions of water of about 40 ml each. Then add 4.0 g of potassium cyanide, KCN, (CAUTION!) to the solution, dissolve it and make up to volume. Mix thoroughly.
(5) Indicator. Mix 20 g of sodium chloride, NaCl, with 100 mg of 2-hydroxy-1-(2-hydroxy-4-sulpho-1-naphtylazo)-3-naphtoic acid, $C_{21}H_{14}N_2O_7S$, (abbreviated to HHSNN).
(6) EDTA solution, 0.0025 mol/l. Dissolve 0.93 g of ethylene diamine tetra acetic acid disodium salt, Na_2EDTA, in 1 litre of water. Standardize according to the procedure outlined below.
(7) Standard solution, Ca concentration 0.00250 mol/l. Transfer 0.2502 g of calcium carbonate, $CaCO_3$, with the help of about 200 ml water into a 1000-ml volumetric flask. Add 6 ml of 1 M HCl (1) and boil to expel carbon dioxide. Allow to cool. Add 40 ml of concentrated sulphuric acid (96%), allow to cool again and make up to volume.

Remarks:

- The indicator HHSNN is also known as Cal-red and as Calcon-carboxylic acid.
- Application of polyvinyl alcohol also prevents the formation of precipitates, but in this case no colour change is shown anymore.
- If Nymcel ZLF 20 is not available, Nymcel ZYF 30 can be used (to be purchased from Nyma, P.O. Box 31, 6500 AA Nijmegen, The Netherlands).

Standard series

With this titration no standard series is required, but the EDTA solution (6) should be standardized against the calcium standard solution (7) in exactly the same way as the sample digests are titrated.

Procedure

(a) Standardization of EDTA

Pipette A ml ($5.00 < A \leq 20.00$, depending on the capacity of the burette) of the calcium standard solution (7) and titrate it following procedure (b). Perform this titration in triplicate.

Calculate the titer of the EDTA solution (6) by

$$t = 0.00250 \times \frac{A}{p}$$

in which

t = titer (= concentration) of the EDTA solution, in mol/l;
A = volume of calcium standard solution (7) pipetted, in ml;
p = average volume of EDTA solution used for titration of the standard solution, in ml;

(b) Titration of digests

Pipette D ml (D varying between 1 and 20, depending on the expected amount of calcium, which should be at most 1.5 mg) of the sample digest into a 100-ml beaker. Put the beaker on a magnetic stirrer and insert a magnetic stirring bar. Start stirring and continue it until the end of the titration.

Add 4.0 ml of magnesium additive (3) and a very small knife-point (10 mg) of indicator (5). *Note: in the still acidic medium the indicator hardly dissolves. Therefore stop the addition of indicator when a faint red colour is observed.* Add dropwise potassium hydroxide solution (2) to neutralize the acid and to dissolve the indicator. Proceed until the colour changes from faint red via practically colourless or slightly yellow to blue (pH 9–10). Then add 1 ml of masking solution (4), followed by 4 ml of potassium hydroxide solution (2); the solution should now show a red colour. Add water until the total volume of liquid in the beaker is about 40 ml.

Titrate with standardized 0.0025 M EDTA solution (6) until the indicator does not show a red hue anymore. The colour should now be blue, but because of the masking discolourations might show.

Titrate a blank digest in an identical manner.

Record the burette readings to the nearest 0.01 ml.

Calculation

The calcium content of the dried plant material, expressed in mmol/kg Ca, is calculated by

$$1000 \times (a - b) \times \frac{t}{D} \times \frac{V}{w}$$

in which
a = volume of EDTA solution used for titrating the sample digest, in ml;
b = volume of EDTA solution used for titrating the blank digest, in ml;
t = titer of the EDTA solution, in mol/l;
D = volume of digest taken for titration, in ml;
V = total volume of digest at the end of the digestion procedure, in ml;
w = weight of plant material sample, in g.

5.2.1 DETERMINATION OF **CADMIUM** BY FLAME AAS

Field of application

This determination is intended to be carried out on digest 5.0 (dry-ashing followed by HF treatment).

The cadmium content of the dried plant material should be higher than 5 mg/kg.

Samples with lower Cd levels should be determined by method 5.2.2; since this method is more cumbersome, it is recommended to apply method 5.2.1 first to all samples as a screening.

Principle

The sample is nebulized into an air-acetylene flame, where it is vaporized; cadmium compounds are atomized and the cadmium atoms thus formed absorb radiation from a hollow-cathode lamp. The absorption is measured at a wavelength of 228.8 nm.

Remarks:

- The plant material digest appears to create a continuous background absorption, so that background correction is necessary.
- Extremely high Fe levels (> 5 g/kg dry material) will cause noticeable spectral interference because of a structured background. When using deuterium background correction, this may lead to too low analytical results (see literature reference).

Apparatus

- Atomic absorption spectrophotometer, with a device for correcting or measuring background absorption.

Reagents

(1) Hydrochloric acid, c(HCl) = 1.2 mol/l. Add 100 ml of concentrated hydrochloric acid (36%) to about 400 ml water and make up to 1 litre.

(2) Stock solution, Cd concentration 1000 mg/l. Dissolve 2.744 g of cadmium nitrate tetrahydrate, $Cd(NO_3)_2 \cdot 4H_2O$ (standardized according to App.

A-16) in some water in a 1000-ml volumetric flask. Add 10 ml of concentrated hydrochloric acid (36%) and make up to the mark.

(3) Standard solution, Cd concentration 10 mg/l. Pipette 1.00 ml of the stock solution (2) into a 100-ml volumetric flask and make up to the mark.

Remarks:

- Reagents and solutions must be prepared with *freshly* delivered water, since polythene and other plastic vessels may release cadmium on standing. The use of extra purified water is not necessary.
- Frequently, Cd determinations go along with Pb determinations. If so, a mixed standard solution may be prepared with 10 mg/l Cd and 100 mg/l Pb in 0.12 M hydrochloric acid.

Standard series

Pipette 0–0.50–1.0–2.0–3.0–4.0 ml of the standard solution (3) into 100-ml volumetric flasks. Add 70 ml of water to each flask. Then add 10 ml of 1.2 M hydrochloric acid (1); mix well. Make up to volume. This standard series has Cd concentrations of 0–0.05–0.1–0.2–0.3–0.4 mg/l.

Procedure

Nebulize the sample digests, the blank digests and the standard series into a just blue (stoichiometric) air-acetylene flame and measure the absorbance at 228.8 nm. Plot a calibration curve and read the cadmium concentrations.

Remarks:

- The calibration curve is slightly bent towards the x-axis.
- Background correction should be applied.
- Samples with very high Fe contents will give erroneous results. The iron should be removed (e.g. by extraction with acetylacetone) or standard addition should be applied; alternatively a special background correction system, e.g. Smith-Hieftje, should be used.

Calculation

The cadmium content of the dried plant material, expressed in mg/kg Cd, is calculated by

$$(a - b) \times \frac{V}{w}$$

in which

a = concentration of cadmium in the sample digest, in mg/l;
b = concentration of cadmium in the blank digest, in mg/l;
V = total volume of digest at the end of the digestion procedure, in ml;
w = weight of plant material sample, in g.

REFERENCES

J.J. van der Lee, E. Temminghoff, V.J.G. Houba and I. Novozamsky. Background corrections in the determination of Cd and Pb by flame AAS in plant and soil samples with high Fe levels. Appl. Spectrosc. *41* (1987) 388–390.

5.2.2 DETERMINATION OF **CADMIUM** BY FURNACE AAS

Field of application

This determination is intended to be carried out on digest 5.0 (dry-ashing followed by HF treatment).

It is intended for samples that are very low in Cd (< 5 mg/kg); the cadmium content should, however, be higher than [REM]Typesetter please see hardcopy page 9 for number[REM\] μg/kg.

Principle

The sample is brought into a graphite furnace, where it is subsequently dried, ashed and vaporized by electrical heating. The cadmium atoms thus formed absorb radiation from a hollow-cathode lamp. The absorbance is measured at a wavelength of 228.8 nm while applying Zeeman background correction.

Remark:

- When using Zeeman background correction, no extraction with an organic sequestering agent into xylene is needed to eliminate matrix interferences.

Apparatus

- Atomic absorption spectrophotometer, with electrothermal atomization device (graphite furnace) and with a device for correcting or measuring background absorption, preferably a Zeeman background correction system.

Reagents

(1) Hydrochloric acid, c(HCl) = 1.2 mol/l. Add 100 ml of concentrated hydrochloric acid (36%) to about 400 ml water and make up to 1 litre.
(2) Hydrochloric acid, c(HCl) = 0.12 mol/l. Dilute the 1.2 M hydrochloric acid (1) 1 + 9 (v/v).
(3) 1-Butanol (= n-butylalcohol).
(4) Stock solution, Cd concentration 1000 mg/l. Dissolve 2.7442 g of cadmium nitrate tetrahydrate, $Cd(NO_3)_2 \cdot 4H_2O$, (standardized according to App. A-16) in some water in a 1000-ml volumetric flask. Add 10 ml of concentrated hydrochloric acid (36%), and make up to the mark.

(5) Standard solution, Cd concentration 10 mg/l. Pipette 1.00 ml of the stock solution (4) into a 100-ml volumetric flask; make up to the mark with 0.12 M hydrochloric acid (2).

(6) Diluted standard solution, Cd concentration 50 µg/l. Pipette 5.00 ml of the standard solution (5) into a 1000-ml volumetric flask. Add 100 ml of 1.2 M hydrochloric acid (1) and make up to volume.

Remarks:

- Reagents and solutions must be prepared with *freshly* delivered water, since polythene and other plastic vessels may release cadmium on standing. The use of extra purified water is not necessary.
- Frequently, Cd determinations go along with Pb determinations. If so, a mixed diluted standard solution may be prepared with 50 µg/l Cd and 1000 µg/l Pb in 0.12 M hydrochloric acid.

Standard series

Pipette 0–0.50–1.00–1.50–2.00 ml of the diluted standard solution (6) into 100-ml erlenmeyer flasks. Dilute to about 30 ml with 0.12 M hydrochloric acid (2); mix immediately after addition of acid.

Procedure

Transfer (part of) the digests and the standard series to polythene cups that fit in the automatic sampler of the atomic absorption spectrophotometer. Add 0.20 ml of 1-butanol (3) and mix thoroughly with an electric mini-stirrer. Heat the extract in a graphite furnace according to an appropriate time-temperature programme (see remark). Measure the absorbance at 228.8 nm in the atomization phase, in the cleaning phase, and at room temperature at the end of the cycle; use Zeeman background correction. Plot a calibration curve and read the cadmium concentrations.

Remarks:

- When a sample gives a higher absorbance than the highest standard, it should be diluted with the zero standard solution (dilution factor = f); this can be programmed in the automatic sampler.
- The calibration curve is slightly bending towards the x-axis.
- Samples with high Fe content may give too low results when applying a deuterium background correction system.
- The absorbance reading in the cleaning phase is meant as a check on memory effects. If these are observed, the results of the last and the second last sample should be discarded. With the programme used in our laboratory (see remark), memory effects are only observed if something is out of order.
- After completion of each firing, the absorbance is read at room temperature as an indication of baseline drift. This value is used as a correction on the reading in the atomization phase.
- Every sample should be measured at least three times, since the precision of this technique is low.
- The wash solution of the automatic sampler contains 2-propanol (2% in water) in order to lower its surface tension.
- The measurements can be performed with any ETA-AAS system. The present method was worked out using a Varian SpectrAA 300 system. The optimum programme – after 30 min warming up – is as follows:

phase	I (dry)	II (char)	III (atomize)	IV (clean)	"V" (base line)
temperature (°C)	160	450	2200	2500	ambient
rate	1	0	1	0	—
hold time (s)	30	45	10	3	—
gas stop (s)	—	—	10	—	—
read	—	—	yes	yes	yes

- When using other instruments, the optimum temperature values to be set may differ from the values given above, because the temperatures mentioned here are instrument settings instead of real temperature values.

Calculation

The cadmium content of the dried plant material, expressed in µg/kg Cd, is calculated by

$$\frac{(a - b) \times f \times V}{w}$$

in which

a = concentration of cadmium in the sample digest, in µg/l;
b = concentration of cadmium in the blank digest, in µg/l;
f = dilution factor for the measuring solution;
V = total volume of digest at the end of the digestion procedure, in ml;
w = weight of plant material sample, in g.

5.2.3 DETERMINATION OF **CADMIUM** BY FURNACE AAS AFTER LIQUID-LIQUID EXTRACTION

Field of application

This determination is intended to be carried out on digest 5.0 (dry-ashing followed by HF treatment).

It is intended for samples that are very low in Cd (< 5 mg/kg); the cadmium content should, however, be higher than 25 µg/kg.

Method 5.2.3 should only be applied when no Zeeman background correction system is available.

Principle

Cadmium ions are sequestered by a carbamate, and the sample is extracted into xylene. The extract is subsequently dried, ashed and vaporized by electrical heating in a graphite furnace. The cadmium atoms thus formed absorb radiation from a hollow-cathode lamp. The absorbance is measured at a wavelength of 228.8 nm.

Remarks:

- The extraction serves two purposes at a time: a large part of the matrix interferences is eliminated, and the cadmium is more concentrated.
- Since not all interferences will have disappeared, background correction is necessary.

Apparatus

- Erlenmeyer flasks with ground glass necks and PTFE stoppers.
- Linear shaker (see Fig. 1).
- Water repellent filter paper.
- Polythene cups.
- Atomic absorption spectrophotometer, with electrothermal atomization device (graphite furnace) and with a device for correcting or measuring background absorption.

Reagents

(1) Hydrochloric acid, c(HCl) = 1.2 mol/l. Add 100 ml of concentrated hydrochloric acid (36%) to about 400 ml water and make up to 1 litre.
(2) Hydrochloric acid, c(HCl) = 0.12 mol/l. Dilute the 1.2 M hydrochloric acid (3) 1 + 9 (v/v).
(3) DDDTC solution. Dissolve 5 g of diethylammonium-N,N-diethyldithiocarbamate, $C_9H_{22}N_2S_2$, in 100 ml xylene. Stir with a magnetic stirrer until all crystals are dissolved; this will take about 45 minutes. The solution can be kept overnight at room temperature.
(4) 1-Butanol (= n-butylalcohol).
(5) Stock solution, Cd concentration 1000 mg/l. Dissolve 2.7442 g of cadmium nitrate tetrahydrate, $Cd(NO_3)_2 \cdot 4H_2O$, (standardized according to App. A-16) in some water in a 1000-ml volumetric flask. Add 10 ml of concentrated hydrochloric acid (36%), and make up to the mark.
(6) Standard solution, Cd concentration 10 mg/l. Pipette 1.00 ml of the stock solution (5) into a 100-ml volumetric flask; make up to the mark with 0.12 M hydrochloric acid (2).
(7) Diluted standard solution, Cd concentration 50 μg/l. Pipette 5.00 ml of the standard solution (6) into a 1000-ml volumetric flask. Add 100 ml of 1.2 M hydrochloric acid (1) and make up to volume.

Remarks:

- Reagents and solutions must be prepared with *freshly* delivered water, since polythene and other plastic vessels may release cadmium on standing. The use of extra purified water is not necessary.
- Frequently, Cd determinations go along with Pb determinations. If so, a mixed diluted standard solution may be prepared with 50 μg/l Cd and 1000 μg/l Pb in 0.12 M hydrochloric acid.
- The solid DDDTC should be stored in a refrigerator in order to prevent deterioration.

Standard series

Pipette 0–0.50–1.00–1.50–2.00 ml of the diluted standard solution (7) into 100-ml erlenmeyer flasks. If Pb is also to be determined, pipette 0–0.50–1.00–1.50–2.00 ml of a 1000 µg/l Pb solution (see 5.8.3) into the same erlenmeyer flasks. Dilute to about 30 ml with 0.12 M hydrochloric acid (2); mix immediately after addition of acid.

Procedure

(a) Pipetting of digests
Pipette D ml (D varying between 1 and 30 ml, according to the expected cadmium concentrations) of the sample digests into 100-ml erlenmeyer flasks; pipette also E ml (E = highest D) of the blank digests into 100-ml erlenmeyer flasks. Dilute to 30 ml with 0.12 M hydrochloric acid (2).

(b) Extraction
Add both to standard series, sample digests and blank digests 10.00 ml of the DDDTC solution (3) and close the erlenmeyer flasks with a PTFE stopper. Shake mechanically (see Fig. 1) during 4 min (stroke length 9 cm, 150 oscillations/min). Suck off the water phase (see Fig. 2) to such an extent that the remaining water will not occupy more than half of the filter area, and filter over water repellent filter paper.

Shake an extra amount of five portions of the zero standard for future dilution purposes; mix the five extracts to make a uniform zero standard (pool).

After the extraction, the standard series has cadmium concentrations of 0–2.5–5–7.5–10 µg/l. (If lead was also put in, its concentrations are 0–50–100–150–200 µg/l).

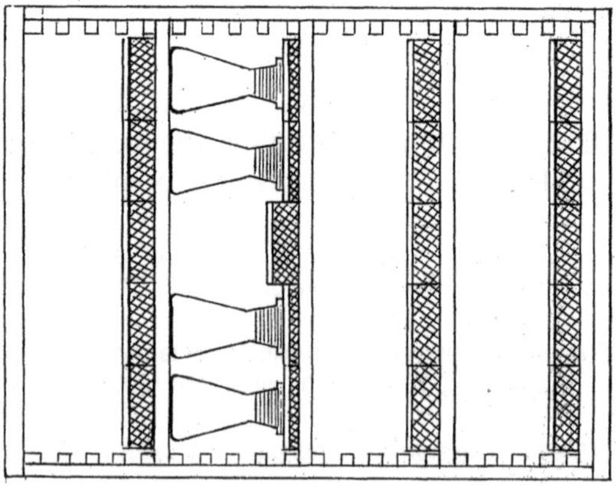

Fig. 1. Rack & erlenmeyer flasks (top view).

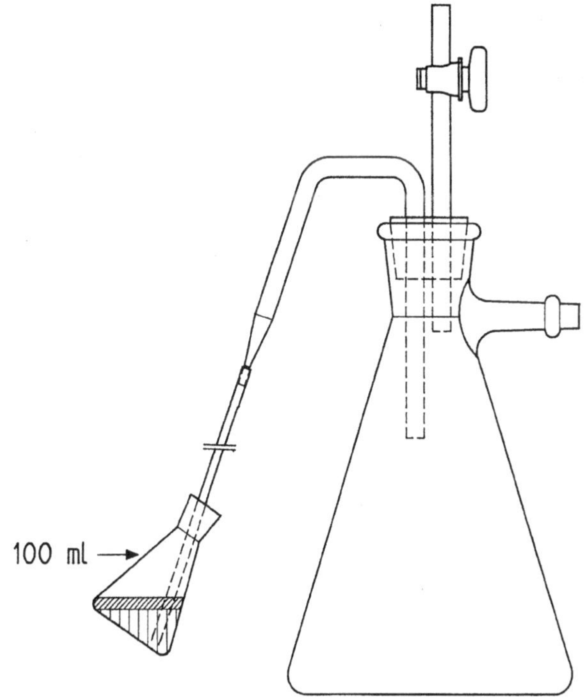

Fig. 2. Sucking off the (lower) water phase.

(c) Measurement

Pipette 1.00 ml of the extracts obtained at (b) into polythene cups that fit in the automatic sampler of the atomic absorption spectrophotometer. Add 0.20 ml of 1-butanol (4) and mix thoroughly with an electric mini-stirrer.

Heat the extract in a graphite furnace according to an appropriate time-temperature programme (see remark). Measure the absorbance at 228.8 nm in the atomization phase, in the cleaning phase, and at room temperature at the end of the cycle; use background correction.

Plot a calibration curve and read the cadmium concentrations.

Remarks:

- Samples and standard series should be shaken at one time, or at least directly after each other, with the same shaking machine.
- When a sample gives a higher absorbance than the highest standard, it should be diluted with the extract of the zero standard solution (dilution factor = f). If a more than tenfold dilution would be required, then the extraction should be repeated with less digest pipetted (D ml).
- The calibration curve is slightly bending towards the x-axis.
- Xylene vapours are rather poisonous. Work in a fume hood.
- For samples with high Fe content, an appreciable amount of Fe can be transferred into the organic phase. This may give too low results when applying a deuterium background correction system.
- The absorbance reading in the cleaning phase is meant as a check on memory effects. If these are observed, the results of the last and the second last sample should be discarded.

With the programme used in our laboratory (see remark), memory effects are only observed if something is out of order.

- After completion of each firing, the absorbance is read at room temperature as an indication of baseline drift. This value is used as a correction on the reading in the atomization phase.
- Every sample should be measured at least three times, since the precision of this technique is low.
- Clean all glassware, polythene cups and PTFE stoppers as follows: After use, wash once with ethanol 96% to remove the remaining xylene. Let drain, rinse with water and shake with 4 M HNO_3 for 5 minutes (or leave overnight). Then rinse with water and twice with ethanol 96%. Allow to dry by leaving at room temperature in an inverted position.
- The wash solution of the automatic sampler contains 2-propanol (2% in water) in order to lower its surface tension.
- The complex of Cd-DDDTC in xylene is stable for about 24 hours, and can therefore be measured on "extraction day" or the day after.
- The measurements can be performed with any ETA-AAS system. The present method was worked out using a Perkin-Elmer system, consisting of a model 460 AA spectrophotometer, a HGA 76 B graphite furnace, an AS-1 automatic injection unit annex sampler and a PRS-10 printer. With a lamp current of 6 mA, a gain of 40, an integration time of 10 s, background correction on (mode AA minus BG), slit 0.7 nm Alt and peak height mode, the optimum programme – after 30 min warming up – is as follows:

phase	I (dry)	II (char)	III (atomize)	IV (clean)	"V" (base line)
temperature (°C)	160	450	2200	2500	ambient
rate	1	0	1	0	—
hold time (s)	30	45	10	3	—
gas stop (s)	—	—	10	—	—
read	—	—	yes	yes	yes

- When using other instruments, the optimum temperature values to be set may differ from the values given above, because the temperatures mentioned here are instrument settings instead of real temperature values.

Calculation

The cadmium content of the dried plant material, expressed in µg/kg Cd, is calculated by

$$\left\{\frac{E \times a}{D} - b\right\} \times \frac{10 \times f \times V}{E \times w}$$

in which

a = concentration of cadmium in the xylene extract of the sample digest, in µg/l;

b = concentration of cadmium in the xylene extract of the blank digest, in µg/l;

f = dilution factor for the measuring solution (remark);

V = total volume of digest at the end of the digestion procedure, in ml;

D = volume of sample digest pipetted for extraction, in ml;

E = volume of blank digest pipetted for extraction, in ml; (E = highest D)

w = weight of plant material sample, in g.

Remark:
- If f = 1, D = 30, V = 100 and w = 2, then the cadmium content is (a — b)/60.

5.3.1 DETERMINATION OF **COPPER** BY FLAME AAS

Field of application

This determination is intended to be carried out on digest 5.0 (dry-ashing followed by HF treatment).

Samples which are expected to be very low in Cu (< 25 mg/kg) should be determined by method 5.3.2; the present flame AAS method is especially suited for screening a series of samples in order to see which samples must be analysed by (the more cumbersome) method 5.3.2.

Principle

The sample is nebulized into an air-acetylene flame, where it is vaporized; copper compounds are atomized and the copper atoms thus formed absorb radiation from a hollow-cathode lamp. The absorbance is measured at a wavelength of 324.7 nm.

Apparatus

- Atomic absorption spectrophotometer.

Reagents

(1) Hydrochloric acid, c(HCl) = 0.12 mol/l. Add 10 ml of concentrated hydrochloric acid (36%) to about 400 ml water and make up to 1 litre.
(2) Stock solution, Cu concentration 1000 mg/l. Dissolve 3.929 g of copper(II) sulphate pentahydrate, $CuSO_4 \cdot 5H_2O$ (standardized according to Appendix A-18) in a 1000-ml volumetric flask in 0.12 M hydrochloric acid (1), and make up to the mark with this acid.
(3) Standard solution, Cu concentration 50 mg/l. Pipette 25.0 ml of the stock solution (2) into a 500-ml volumetric flask and make up to volume with 0.12 M hydrochloric acid (1).

Remarks:
- Frequently, Cu determinations go along with Fe determinations. In that case, a mixed standard solution may be prepared with 50 mg/l Cu and 50 mg/l Fe (see 5.4.1).

Standard series

Pipette 0–2.00–4.00–6.00–8.00–10.00 ml of the standard solution (3) into 100-ml volumetric flasks and make up to volume with 0.12 M hydrochloric acid

(1). This standard series has Cu concentrations of 0–1–2–3–4–5 mg/l.

Procedure

Nebulize the sample digests, the blank digests and the standard series into a just blue (stoichiometric) air-acetylene flame and measure the absorbance at a wavelength of 324.7 nm. Plot a calibration curve and read the copper concentrations.

Remarks:
- The calibration curve is slightly bent towards the x-axis, therefore the curve should not be generated by means of linear regression.
- It is not necessary to apply background correction.
- No interferences have been found.

Calculation

The copper content of the dried plant material, expressed in mg/kg Cu, is calculated by

$$(a - b) \times \frac{V}{w}$$

in which
a = concentration of copper in the sample digest, in mg/l;
b = concentration of copper in the blank digest, in mg/l;
V = total volume of digest at the end of the digestion procedure, in ml;
w = weight of plant material sample, in g.

5.3.2 DETERMINATION OF **COPPER** BY FURNACE AAS

Field of application

This determination is intended to be carried out on digest 5.0 (dry-ashing followed by HF treatment).

This method is especially suited for low copper contents; since it is rather time-consuming, a preliminary screening by flame AAS (section 5.3.1) is recommended.

Principle

The sample is brought into a graphite furnace, where it is subsequently dried, ashed and vaporized by electrical heating. The copper atoms thus formed absorb radiation from a hollow-cathode lamp. The absorbance is measured at a wavelength of 324.7 nm while applying Zeeman background correction.

Remarks:
- When using Zeeman background correction, no extraction with an organic sequestering agent into xylene is needed to eliminate matrix interferences.

Apparatus

- Atomic absorption spectrophotometer, with electrothermal atomization device (graphite furnace) and with a device for correcting or measuring background absorption, preferably a Zeeman background correction system.

Reagents

(1) Hydrochloric acid, c(HCl) = 1.2 mol/l. Add 100 ml of concentrated hydrochloric acid (36%) to about 400 ml water and make up to 1 litre.
(2) Hydrochloric acid, c(HCl) = 0.12 mol/l. Dilute the 1.2 M hydrochloric acid (1) 1 + 9 (v/v).
(3) 1-Butanol (= n-butylalcohol).
(4) Stock solution, Cu concentration 1000 mg/l. Dissolve 3.929 g of copper(II) sulphate pentahydrate, $CuSO_4 \cdot 5H_2O$ (standardized according to App. A-18) in about 100 ml water in a 1000-ml volumetric flask. Add 100 ml of 1.2 M hydrochloric acid (1) and dilute to volume.
(5) Standard solution, Cu concentration 10 mg/l. Pipette 1.00 ml of the stock solution (4) into a 100-ml volumetric flask and make up to the mark with 0.12 M hydrochloric acid (2).
(6) Diluted standard solution, Cu concentration 50 µg/l. Pipette 5.00 ml of the standard solution (5) into a 1000-ml volumetric flask. Add 100 ml of 1.2 M hydrochloric acid (1) and make up to volume.

Remark:
- Frequently, Cu determinations go along with Fe determinations. If so, a mixed diluted standard solution may be prepared with 50 µg/l Cu and 50 µg/l Fe in 0.12 M hydrochloric acid.

Standard series

Pipette 0–0.50–1.00–1.50–2.00 ml of the diluted standard solution (6) into 100-ml erlenmeyer flasks. Dilute to about 30 ml with 0.12 M hydrochloric acid (2); mix immediately after addition of acid.

Procedure

Transfer (part of) the digests and the standard series to polythene cups that fit in the automatic sampler of the atomic absorption spectrophotometer. Add 0.20 ml of 1-butanol (3) and mix thoroughly with an electric mini-stirrer. Heat the extract in a graphite furnace according to an appropriate time-temperature programme (see remark). Measure the absorbance at 324.7 nm in the atomization phase, in the cleaning phase, and at room temperature at the end of the cycle; use Zeeman background correction. Plot a calibration curve and read the copper concentrations.

Remarks:

- When a sample gives a higher absorbance than the highest standard, it should be diluted with the zero standard solution (dilution factor = f); this can be programmed in the automatic sampler.
- The calibration curve is slightly bending towards the x-axis.
- The absorbance reading in the cleaning phase is meant as a check on memory effects. If these are observed, the results of the last and the second last sample should be discarded. With the programme used in our laboratory (see remark), memory effects are only observed if something is out of order.
- After completion of each firing, the absorbance is read at room temperature as an indication of baseline drift. This value is used as a correction on the reading in the atomization phase.
- Every sample should be measured at least three times, since the precision of this technique is low.
- The wash solution of the automatic sampler contains 2-propanol (2% in water) in order to lower its surface tension.
- The measurements can be performed with any ETA-AAS system. The present method was worked out using a Varian SpectrAA 300 system. The optimum programme – after 30 min warming up – is as follows:

phase	I (dry)	II (char)	III (atomize)	IV (clean)	"V" (base line)
temperature (°C)	160	450	2200	2500	ambient
rate	1	0	1	0	—
hold time (s)	30	45	10	3	—
gas stop (s)	—	—	10	—	—
read	—	—	yes	yes	yes

- When using other instruments, the optimum temperature values to be set may differ from the values given above, because the temperatures mentioned here are instrument settings instead of real temperature values.

Calculation

The copper content of the dried plant material, expressed in µg/kg Cu, is calculated by

$$\frac{(a - b) \times f \times V}{w}$$

in which

a = concentration of copper in the sample digest, in µg/l;
b = concentration of copper in the blank digest, in µg/l;
f = dilution factor for the measuring solution;
V = total volume of digest at the end of the digestion procedure, in ml;
w = weight of plant material sample, in g.

5.3.3 DETERMINATION OF **COPPER** BY COLORIMETRY AFTER LIQUID-LIQUID EXTRACTION

Field of application

This determination is intended to be carried out on digest 5.0 (dry-ashing followed by HF treatment).

This method is especially suited for low copper contents; since it is rather time-consuming, a preliminary screening by flame AAS (section 5.3.1) is recommended.

Principle

Copper(II) ions are sequestered by a carbamate; the yellow-coloured complex is extracted from sulphuric acid medium into xylene. Its absorbance is measured at a wavelength of 435 nm.

Remarks:

- The extraction is intended to concentrate the copper, since its content in plant material is normally very low. At the same time, however, the usually interfering metals are not extracted into the organic phase because of the low pH.
- The sulphuric acid concentration should be between 0.3–1.1 mol/l.

Apparatus

- Linear shaker.
- Rack for erlenmeyer flasks to be placed in the shaker (see Fig. 3).
- Water repellent filter paper (e.g., Machery & Nagel 616 WA, or Schleicher & Schüll 597 hy, or Whatman 1PS).
- Colorimeter.

Reagents

(1) Sulphuric acid, $c(H_2SO_4) = 2.5$ mol/l. Add carefully, while swirling, 140 ml of concentrated sulphuric acid (96%) to about 400 ml water. Allow to cool and make up to 1 litre.

(2) Sulphuric acid, $c(H_2SO_4) = 0.08$ mol/l. Add 4.5 ml of concentrated sulphuric acid (96%) to about 400 ml of water and make up to 1 litre.

(3) Zinc dibenzyldithiocarbamate solution, 0.02% (w/v). Dissolve 100 mg of zinc dibenzyldithiocarbamate in 500 ml of xylene. Prepare fresh weekly.

(4) Stock solution, Cu concentration 1000 mg/l. Dissolve 3.929 g of copper(II) sulphate pentahydrate, $CuSO_4 \cdot 5H_2O$ (standardized according to App. A-18) in about 100 ml water in a 1000-ml volumetric flask. Add 32 ml of 2.5 M sulphuric acid (1) and dilute to volume.

(5) Standard solution, Cu concentration 1 mg/l. Pipette 1.00 ml of the stock solution (4) into a 1000-ml volumetric flask and make up to the mark with 0.08 M sulphuric acid (2).

Remark:

- The commonly used sodium diethyldithiocarbamate gives a less stable complex, and must be used in slightly alkaline medium. Then it forms also complexes with other metals. The zinc dibenzyldithiocarbamate, however, is stable in acid medium.

Standard series

Pipette 0–2.00–4.00–6.00–8.00–10.00 ml of the standard solution (5) into 100-ml erlenmeyer flasks that have ground glass necks and PTFE stoppers. Add respectively 20–18–16–14–12–10 ml of 0.08 M sulphuric acid (2). At this moment, this standard series has Cu concentrations of 0–0.1–0.2–0.3–0.4–0.5 mg/l.

Procedure

Extraction
Pipette D ml (D being mostly 20.00) of the sample digests and the blank digests into 100-ml erlenmeyer flasks that have ground glass necks and PTFE stoppers. Add to the standard series, the sample digests and the blank digests 10 ml of 2.5 M sulphuric acid (1). Add (40-D) ml water and mix. Then add 5.00 ml of the zinc dibenzyldithiocarbamate solution (3). Stopper the erlenmeyer flasks, place them in the rack (see Fig. 3) and shake mechanically during 10 min (stroke length 9 cm, 150 oscillations per min). Wait until the phases have separated. Suck off the water phase (see Fig. 4) to such an extent that the filter to be used can easily retain the remaining water. Filter over water repellent paper into test tubes.

Measurement
Calibrate the spectrophotometer while the cuvette is filled with xylene. Measure the absorbance of all extracts in a 1-cm cuvette at a wavelength of 435 nm. Plot a calibration curve and read the copper concentrations.

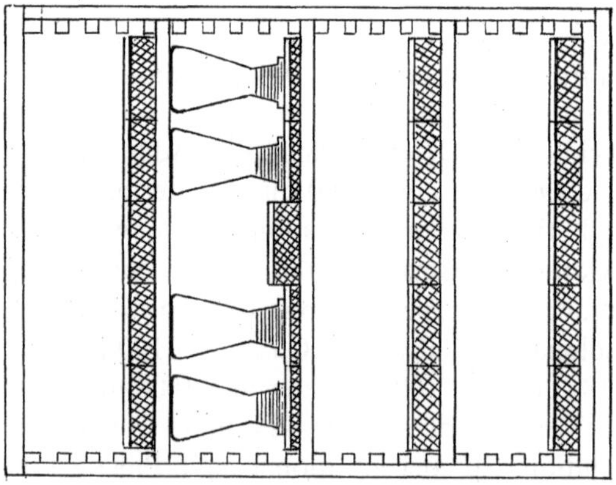

Fig. 3. Rack & erlenmeyer flasks (top view).

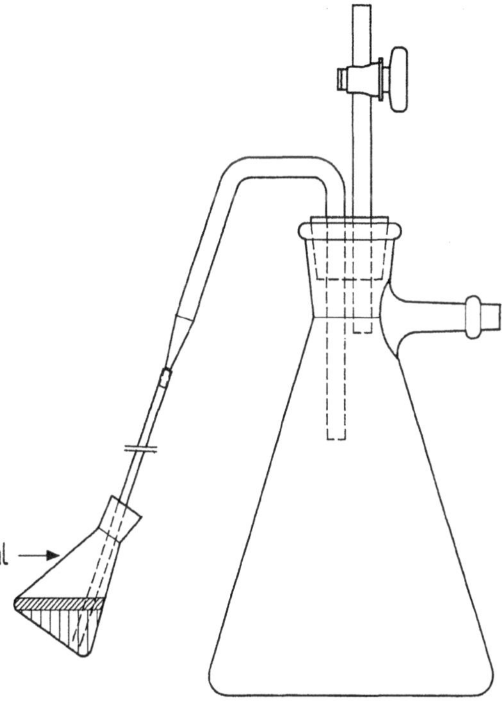

Fig. 4. Sucking off the (lower) water phase.

Remarks:

- Samples and standard series should be shaken together at the same time, on a powerful linear shaker at the prescribed conditions. Otherwise, the extraction yield may be variable and low.
  If the conditions of 9 cm stroke length and/or 150 strokes per min cannot be fulfilled, the time of shaking should be prolonged; the optimum time should be established by shaking standard series for different periods.
- For very low concentrations, 40 ml of digest may be taken for extraction; for relatively high concentrations 5 ml will suffice.
- The colour is stable for at least 5 hours.
- The calibration curve should be linear.
- The erlenmeyer flasks should be cleaned in advance by shaking with a suspension of sodium diethyldithiocarbamate in chloroform (1% w/v). Rinse thoroughly only with demineralised water. (Chloroform and sodium diethyldithiocarbamate are used because the washing is quite easy.)
- Xylene and chloroform vapours are rather poisonous. Work in a fume hood.
- When taking 20 ml of digest 3.0, no interferences were found from concentrations of 100 mg/l Al, 5 mg/l Co, 100 mg/l Fe, 25 mg/l Mn, 10 mg/l Ni and 30 mg/l Zn in the digest. A concentration of 10 mg/l Mo causes a signal enhancement of 3% at a level of 2.5 mg/l Cu.

Calculation

The copper content of the dried plant material, expressed in mg/kg Cu, is calculated by

$$20 \times (a - b) \times \frac{t = 1}{D} \times \frac{V}{w}$$

in which

a = concentration of copper in the xylene extract of the sample digest, in mg/l;
b = concentration of copper in the xylene extract of the blank digest, in mg/l;
D = volume of digest pipetted for extraction, in ml;
V = total volume of digest at the end of the digestion procedure, in ml;
w = weight of plant material sample, in g.

REFERENCES

S. Andrus. Determination of copper in plant materials by means of zinc dibenzyldithiocarbamate. Analyst *80* (1955) 514–516.

R.I. Martens and R.E. Githens, Sr. Small amounts of copper in dyes and rubber chemicals; colorimetric determination with zinc dibenzyldithiocarbamate. Anal. Chem. *24* (1952) 991–993.

5.4.1 DETERMINATION OF **IRON** BY FLAME AAS

Field of application

This determination is intended to be carried out on digest 5.0 (dry-ashing followed by HF treatment).

The iron content of the dried plant material should be higher than 25 mg/kg.

Principle

The sample is nebulized into an air-acetylene flame, where it is vaporized; iron compounds are atomized and the iron atoms thus formed absorb radiation from a hollow-cathode lamp. The absorbance is measured at a wavelength of 248.3 nm.

Apparatus

- Atomic absorption spectrophotometer.

Reagents

(1) Hydrochloric acid, c(HCl) = 0.12 mol/l. Add 10 ml of concentrated hydrochloric acid (36%) to about 400 ml water and make up to 1 litre.
(2) Stock solution, Fe concentration 1000 mg/l. Put about 200 ml water in a 1000-ml volumetric flask and add 10 ml of concentrated hydrochloric

acid (36%). Weigh out 7.022 g of iron(II) ammonium sulphate hexahydrate, $Fe(NH_4)_2(SO_4)_2 \cdot 6H_2O$, standardized according to Appendix A-19, and transfer it with some water to this 1000-ml volumetric flask. Boil to dissolve the salt completely. Allow to cool and make up to the mark.

(3) Standard solution, Fe concentration 50 mg/l. Pipette 25.0 ml of the stock solution (2) into a 500-ml volumetric flask. Make up to volume with 0.12 M hydrochloric acid (1).

Remark:

- Frequently, Fe determinations go along with Cu determinations. If so, a mixed standard solution may be prepared with 50 mg/l Fe and 50 mg/l Cu (see section 5.3.1).

Standard series

Pipette 0–2.00–4.00–6.00–8.00–10.00 ml of the standard solution (3) into 100-ml volumetric flasks. Make up to volume with 0.12 M hydrochloric acid (1). This standard series has Fe concentrations of 0–1–2–3–4–5 mg/l.

Procedure

Nebulize the sample digests, the blank digests and the standard series into a blue (oxidizing) air-acetylene flame and measure the absorbance at a wavelength of 248.3 nm. Plot a calibration curve and read the iron concentrations.

Remarks:

- The calibration curve is slightly bent towards the x-axis.
- It is not necessary to apply background correction.
- No interferences have been found.

Calculation

The iron content of the dried plant material, expressed in mg/kg Fe, is calculated by

$$(a - b) \times \frac{V}{w}$$

in which
a = concentration of iron in the sample digest, in mg/l;
b = concentration of iron in the blank digest, in mg/l;
V = total volume of digest at the end of the digestion procedure, in ml;
w = weight of plant material sample, in g.

5.4.2 DETERMINATION OF **IRON** BY COLORIMETRY

Field of application

This determination is intended to be carried out on digest 5.0 (dry-ashing followed by HF treatment).

Principle

Iron(II) ions form an orange-red coloured complex with 1,10-phenanthroline at pH 2–9. Its absorbance is measured at a wavelength of 515 nm.

Remarks:

- Metol is used to reduce Fe(III) ions.
- When both aluminium and phosphate are present, the solution may become turbid. To prevent this, citrate is added.

Apparatus

- Colorimeter.

Reagents

(1) Hydrochloric acid, c(HCl) = 0.12 mol/l. Add 10 ml of concentrated hydrochloric acid (36%) to about 400 ml water and make up to 1 litre.
(2) 1,10-Phenanthroline solution. Dissolve 0.40 g of 1,10-phenanthroline hydrochloride hydrate, $C_{12}H_8N_2HCl \cdot H_2O$, in 25 ml of ethanol 96% (v/v). Make up to 100 ml with water.
(3) Citrate solution. Dissolve 2.20 g of sodium citrate dihydrate, $Na_3C_6H_5O_7 \cdot 2H_2O$, in some water and make up to 100 ml.
(4) Metol reductant solution. Dissolve 0.10 g of metol (p-methylamino-phenol sulphate, $C_{14}H_{18}N_2O_2 \cdot H_2SO_4$), 13.7 g of sodium metabisulphite, $Na_2S_2O_5$, and 0.5 g of sodium sulphite, Na_2SO_3, in some water and make up to 100 ml.
(5) Stock solution, Fe concentration 1000 mg/l. Dissolve 7.022 g of iron(II) ammonium sulphate hexahydrate, $Fe(NH_4)_2(SO_4)_2 \cdot 6H_2O$, standardized according to App. A-19, in about 100 ml of water in a 1000-ml volumetric flask. Add 10 ml of concentrated hydrochloric acid (36%). Boil to dissolve the salt completely and allow to cool down. Make up to volume with water.
(6) Standard solution, Fe concentration 100 mg/l. Pipette 50 ml of the stock solution (5) into a 500-ml volumetric flask and make up to the mark with 0.12 M hydrochloric acid (1).

Standard series

Pipette 0–2.00–4.00–6.00–8.00–10.00 ml of the standard solution (6) into 100-ml volumetric flasks and make up to volume with 0.12 M hydrochloric acid (1). This series has Fe concentrations of 0–2–4–6–8–10 mg/l.

Procedure

Pipette 2.00 ml of the sample digests, the blank digests and the standard series into test tubes. Add, in this sequence,

- 0.10 ml of metol reductant solution (4);
- 0.20 ml of 1,10-phenanthroline solution (2);
- 2.70 ml of citrate solution (3).

Mix after each addition. Wait for at least 1 h, then measure the absorbance in a 1-cm cuvette at a wavelength of 515 nm. Plot a calibration curve and read the Fe concentrations.

Remarks:

- The colour intensity is constant for at least 5 h.
- The calibration curve should be linear.
- Some heavy metals form complexes with 1,10-phenanthroline without showing a colour. Therefore, it is advised to do the measurements in duplicate, and to pipette then 1.00 ml instead of 2.00 ml of the digests and to add 1.00 ml of 0.12 M hydrochloric acid (1).
  If the 1-ml aliquots give higher results than the 2-ml aliquots, the measurements should be repeated with 0.5-ml aliquots.
- No interferences were found from concentrations of 2000 mg/l PO_4 and 200 mg/l Al, even if present simultaneously. Furthermore, no interferences were found from 10 mg/l Co, 25 mg/l Cu, 10 mg/l Mo, 200 mg/l Ni and 100 mg/l Zn.

Calculation

The iron content of the dried plant material, expressed in mg/kg Fe, is calculated by

$$(a - b) \times \frac{V}{w}$$

in which
a = concentration of iron in the sample digest, in mg/l;
b = concentration of iron in the blank digest, in mg/l;
V = total volume of digest at the end of the digestion procedure, in ml;
w = weight of plant material sample, in g.

5.5 DETERMINATION OF **POTASSIUM** BY FLAME AES

Field of application

This determination is intended to be carried out on digest 5.0 (dry-ashing followed by HF treatment).

The determination may be done in combination with Na (5.7); however, if Na is also to be measured in digest 5.0, another ionization buffer is needed; see determination 5.7.

Principle

The sample is nebulized into an air-propane flame, where it is vaporized; potassium compounds are atomized and the potassium atoms thus formed emit radiation of which the intensity is measured at a wavelength of 766.5 nm.

Remark:

- To prevent ionisation interferences, a cesium salt is added, to act as an ionisation buffer.

Apparatus

- Flame emission spectrometer.

Reagents

(1) Cesium solution, Cs concentration 0.87 g/l. Dissolve 1.1 g of cesium chloride, CsCl, "pro analysi", in 1 litre of water.

(2) Stock solution, K concentration 4000 mg/l. Dissolve 7.627 g of potassium chloride, KCl, in some water in a 1000-ml volumetric flask and make up to the mark with water.

Remarks:

- The cesium chloride should be of the highest analytical grade ("pro analysi"), because a lower quality (e.g. "reinst" = "most pure") contains much more Na.
- Frequently, K determinations go along with Na determinations. If so, a mixed stock solution may be prepared with 4000 mg/l K and 1250 mg/l Na (see 5.7).

Standard series

Pipette 0–2.00–4.00–6.00–8.00–10.00–12.00 ml of the stock solution (2) into 100-ml volumetric flasks, to which already 50 ml water and 10 ml of concentrated hydrochloric acid (36%) per flask have been added in advance. Make up to the mark with water.

Dilute this standard series 1 + 9 (v/v) with the cesium solution (1); this should be done together with the diluting of the sample digests. This diluted standard series has K concentrations of 0–8–16–24–32–40–48 mg/l.

Procedure

Dilute all digests 1 + 9 (v/v) with the cesium solution (1).

Measure the emission intensity of the diluted sample digests, the diluted blank digests and the diluted standard series at a wavelength of 766.5 nm, using an air-propane flame. Plot a calibration curve and read the potassium concentrations.

Remarks:

- The calibration curve should be nearly linear.
- An air-acetylene flame can be used also, since the Cs concentration is high enough to counteract the greater tendency to ionisation.
- Instead of indicating a wavelength, a simple emission spectrometer ("flame photometer") may only be supplied with a so-called K filter. Make sure that this is an interference filter, since glass filters are not selective enough.
- When using a filter in the spectrometer, the emission from K at the wavelength of 769.9 nm is also measured. This enhances the sensitivity of the K measurement.

Calculation

The potassium content of the dried plant material, expressed in mmol/kg K, is calculated by

$$0.256 \times (a - b) \times \frac{V}{w}$$

in which

a = concentration of potassium in the diluted sample digest, in mg/l;
b = concentration of potassium in the diluted blank digest, in mg/l;
V = total volume of digest at the end of the digestion procedure, in ml;
w = weight of plant material sample, in g.

5.6.1 DETERMINATION OF **MANGANESE** BY FLAME AAS

Field of application

This determination is intended to be carried out on digest 5.0 (dry-ashing followed by HF treatment).

The manganese content of the dried plant material should be higher than 15 mg/kg.

Principle

The sample is nebulized into an air-acetylene flame, where it is vaporized; manganese compounds are atomized and the manganese atoms thus formed absorb radiation from a hollow-cathode lamp. Lanthanum is added to prevent condensed phase interferences. The absorbance is measured at a wavelength of 279.5 nm.

Apparatus

– Atomic absorption spectrophotometer.

Reagents

(1) Hydrochloric acid, c(HCl) = 0.12 mol/l. Add 10 ml of concentrated hydrochloric acid (36%) to about 400 ml water and make up to 1 litre.
(2) Lanthanum solution, La concentration 16.5 g/l. Dissolve 12.86 g of lanthanum nitrate hexahydrate, $La(NO_3)_3 \cdot 6H_2O$ in water and make up to 250 ml.
(3) Stock solution, Mn concentration 500 mg/l. Dissolve 1.438 g of potassium permanganate, $KMnO_4$, (standardized according to Appendix A-21), in about 200 ml of water in a beaker. Add carefully, while stirring, 4.5 ml of concentrated sulphuric acid (96%). Reduce the permanganate with a few drops of hydrogen peroxide (30%) and boil to remove the excess of H_2O_2. Transfer the contents of the beaker quantitatively to a 1000-ml volumetric flask and make up to the mark.
(4) Standard solution, Mn concentration 50 mg/l. Pipette 50.00 ml of the stock solution (3) into a 500-ml volumetric flask and make up to volume with 0.12 M hydrochloric acid (1).

Remarks:
- Care should be taken to match the stock solution (3), the standard solution (4) and the standard series to the acid type and strength of the digest. Thus, for digest 5.0 the medium must be 0.12 M hydrochloric acid, except for the stock solution (3) that must be made in 0.08 M sulphuric acid medium in order to prevent Cl_2 formation.
- Frequently, Mn determinations go along with Zn determinations. If so, a mixed standard solution may be prepared with 25 mg/l Mn and 20 mg/l Zn (see 5.9).

Standard series

Pipette 0–1.00–2.00–4.00–6.00–8.00–10.00 ml of the standard solution (4) into 100-ml volumetric flasks and make up to volume with 0.12 M hydrochloric acid (1). This standard series has Mn concentrations of 0–0.5–1–2–3–4–5 mg/l.

Procedure

Pipette 5.00 ml of the sample digests, the blank digests and the standard series into test tubes. Add 0.50 ml of lanthanum solution (2) and mix. Nebulize all these solutions into a blue (oxidizing) air-acetylene flame and measure the absorbance at 279.5 nm. Plot a calibration curve and read the manganese concentrations.

Remark:
- The calibration curve is slightly bent towards the x-axis, so that calculation by means of linear regression is not allowed.

Calculation

The manganese content of the dried plant material, expressed in mg/kg Mn, is calculated by

$$(a - b) \times \frac{V}{w}$$

in which
a = concentration of manganese in the sample digest, in mg/l;
b = concentration of manganese in the blank digest, in mg/l;
V = total volume of digest at the end of the digestion procedure, in ml;
w = weight of plant material sample, in g.

5.6.2 DETERMINATION OF **MANGANESE** BY COLORIMETRY

Field of application

This determination may be carried out on digest 5.0 (dry-ashing followed by HF treatment).

This determination is only recommended if no atomic absorption spectrophotometer is available.

Principle

In alkaline medium, manganese ions form a strongly coloured complex with formaldoxime. The absorbance of this complex is measured at a wavelength of 500 nm, after having masked any interfering metal ions by KCN.

Remark:

- Fe^{3+} is reduced by metol to Fe^{2+} and then complexed by CN^- to form the less interfering, yellow-coloured $[Fe(CN)_6]^{4-}$. This compound interferes at the wavelength of maximum absorption (450 nm), but not at 500 nm.

Apparatus

- Colorimeter with interference filter.

Reagents

(1) Hydrochloric acid, c(HCl) = 0.12 mol/l. Add 10 ml of concentrated hydrochloric acid (36%) to about 400 ml water and make up to 1 litre.
(2) Metol solution. Transfer 0.10 g of p-methylaminophenolsulphate, $C_{14}H_{20}N_2O_6S$ (metol), 13.7 g of sodium metabisulphite, $Na_2S_2O_5$, and 0.5 g of sodium sulphite, Na_2SO_3, to a 100-ml volumetric flask. Add water and make up to 100 ml. Add a magnetic bar and stir until dissolved.

(3) Cyanide solution. Dissolve 2.0 g of potassium cyanide, KCN, **(CAUTION!)** in 20 ml water, and mix it with 100 ml of concentrated aqueous ammonia (25%). Then add 360 ml water.
(4) Formaldoxime solution. Add 5 g of formaldehyde solution (37%) to 80 ml water. Dissolve 5 g of hydroxylamine-HCl in this mixture. Dilute to 100 ml.
(5) Standard solution, Mn concentration 100 mg/l. Dissolve 0.2877 g of potassium permanganate, $KMnO_4$, (standardized according to App. A-21) in about 200 ml water in a beaker. Add 4.5 ml of concentrated sulphuric acid (96%) and mix. Reduce the permanganate with a few drops of hydrogen peroxide (30%) and boil to remove the excess of H_2O_2. Allow to cool; transfer the contents of the beaker quantitatively to a 1000-ml volumetric flask and make up to the mark.

Remark:

- The standard solution (5) and the standard series should correspond to the acid type and strength of the digest. Thus, for digest 5.0 the medium must be 0.12 M hydrochloric acid, except for the standard solution (5) that should be made in sulphuric acid (0.08 M) in order to avoid Cl_2 formation.

Standard series

Pipette 0–1.00–2.00–4.00–6.00–8.00–10.00 ml of the standard solution (5) into 100-ml volumetric flasks and make up to volume with 0.12 M hydrochloric acid (1). This standard series has Mn concentrations of 0–1–2–4–6–8–10 mg/l.

Procedure

Pipette 2.00 ml of the sample digests, the blank digests and the standard series into test tubes. For duplicate measurements, pipette 1.00 ml of the digests (*not* the standards) into test tubes and add 1.00 ml of 0.12 M hydrochloric acid (1).

Place the racks with the test tubes in the fume cupboard.

Add to all test tubes, in this sequence:

- 0.50 ml metol solution (2) and mix;
- 2.40 ml cyanide solution (3); mix and wait for 15 min;
- 2 drops of formaldoxime solution (4); mix immediately and wait for 30 min.

Measure the absorbance at a wavelength of 500 nm in a 1-cm cuvette. Plot a calibration curve and read the manganese concentrations.

Remarks:

- The cyanide solution is very poisonous. When it is added to the (acidic) digests, HCN may be released, therefore the addition should be done in a fume cupboard. The measurements should be done preferably in the fume cupboard also. After measuring, all CN containing solutions should be collected for appropriate processing.

- If turbidity occurs in the sample digests after the reagents have been added, a knife point of Na_2EDTA may be added. If the turbidity persists, 0.5 ml of sample should be pipetted.
- After the development period, the colour is constant for at least 5 hours.
- The calibration curve should be linear.

Calculation

The manganese content of the dried plant material, expressed in mg/kg Mn, is calculated by

$$(a - b) \times f \times \frac{V}{w}$$

in which
a = concentration of manganese in the sample digest, in mg/l;
b = concentration of manganese in the blank digest, in mg/l;
f = dilution factor: f = 2 when 1 ml was pipetted; f = 1 when 2 ml were pipetted;
V = total volume of digest at the end of the digestion procedure, in ml;
w = weight of plant material sample, in g.

REFERENCES

J.Ch. van Schouwenburg. The routine determination of manganese with formaldoxime. Neth. J. agric. Sci. *14* (1966) 127–130.

5.7 DETERMINATION OF **SODIUM** BY FLAME AES

Field of application

This determination is intended to be carried out on digest 5.0 (dry-ashing followed by HF treatment).

The determination may be done in combination with K (5.5).

Principle

The sample is nebulized into an air-propane flame, where it is vaporized; sodium compounds are atomized and the sodium atoms thus formed emit radiation of which the intensity is measured at a wavelength of 589.0 nm.

Remarks:
- To prevent ionisation interferences, a cesium salt is added to act as an ionization buffer.
- Since the sodium atomic emission line finds itself in the wavelength region of a calcium oxide emission band, too high results will be found if calcium is present in the digest; this is in particular true for digest 5.0. In order to suppress such a matrix interference in HCl

containing solutions, an aluminium salt is added which leads to formation of a poorly dissociated Ca-Al compound.

Apparatus

– Flame emission spectrometer.

Reagents

(1) Cesium-aluminium solution, Cs concentration 0.86 g/l and Al concentration 2.0 g/l. Dissolve 1.1 g of cesium chloride, CsCl, "pro analysi", and 27.7 g of aluminium nitrate nonahydrate, $Al(NO_3)_3 \cdot 9H_2O$, in 1 litre of water.

(2) Stock solution, Na concentration 1250 mg/l. Dissolve 3.178 g of sodium chloride, NaCl, pretreated according to Appendix A-4, in some water in a 1000-ml volumetric flask and make up to the mark with water.

Remarks:

- The cesium chloride should be of the highest analytical grade ("pro analysi"), because a lower quality (e.g. "reinst" = "most pure") contains much more Na.
- Frequently, Na determinations go along with K determinations. If so, a mixed stock solution may be prepared with 1250 mg/l Na and 4000 mg/l K (see 5.5).
- The cesium-aluminium solution (1) must be used for digest 5.0.

Standard series

Pipette 0–2.00–4.00–6.00–8.00–10.00–12.00 ml of the stock solution (2) into 100-ml volumetric flasks, to which already 50 ml water and 10 ml of concentrated hydrochloric acid acid (36%) have been added in advance. Make up to the mark with water. Dilute this standard series 1 + 9 (v/v) with the cesium-aluminium solution (1); this should be done together with the diluting of the sample digests. This diluted standard series has Na concentrations of 0–2.5–5–7.5–10–12.5–15 mg/l.

Procedure

Dilute all digests 1 + 9 (v/v) with the cesium-aluminium solution (1). Measure the emission intensity of the diluted sample digests, the diluted blank digests and the diluted standard series at a wavelength of 589.0 nm, using an air-propane flame. Plot a calibration curve and read the sodium concentrations.

Remarks:

- The calibration curve should be nearly linear.
- An air-acetylene flame can be used also, since the Cs concentration is high enough to counteract the greater tendency to ionisation.
- Instead of indicating a wavelength, a simple emission spectrometer ("flame photometer") may only be supplied with a so-called Na filter. Make sure that this is an interference filter, since glass filters are not selective enough.

Calculation

The sodium content of the dried plant material, expressed in mmol/kg Na, is calculated by

$$0.435 \times (a - b) \times \frac{V}{w}$$

in which
a = concentration of sodium in the diluted sample digest, in mg/l;
b = concentration of sodium in the diluted blank digest, in mg/l;
V = total volume of digest at the end of the digestion procedure, in ml;
w = weight of plant material sample, in g.

5.8.1 DETERMINATION OF **LEAD** BY FLAME AAS

Field of application

This determination is intended to be carried out on digest 5.0 (dry-ashing followed by HF treatment).

The lead content of the dried plant material should be higher than 25 mg/kg.

Samples with lower Pb levels should be determined by method 5.8.2; since this method is more cumbersome, it is recommended to apply method 5.8.1 first to all samples as a screening.

Principle

The sample is nebulized into an air-acetylene flame, where it is vaporized; lead compounds are atomized and the lead atoms thus formed absorb radiation from a hollow-cathode lamp. The absorbance is measured at a wavelength of 217.0 nm.

Remarks:
- The plant material digest appears to create a continuous background absorption, so that background correction is necessary.
- Extremely high Fe levels (> 5 g/kg dry material) will cause noticeable spectral interference because of a structured background. When using deuterium background correction, this may lead to too low analytical results (see literature reference).

Apparatus

- Atomic absorption spectrophotometer, with a device for correcting or measuring background absorption.

Reagents

(1) Hydrochloric acid, c(HCl) = 1.2 mol/l. Add 100 ml of concentrated hydrochloric acid (36%) to about 400 ml water and make up to 1 litre.
(2) Stock solution, Pb concentration 1000 mg/l. Dissolve 1.5985 g of lead nitrate, $Pb(NO_3)_2$ (standardized according to Appendix A-23) in some water in a 1000-ml volumetric flask. Add 10 ml of concentrated nitric acid (65%) and make up to the mark.
(3) Standard solution, Pb concentration 100 mg/l. Pipette 10.0 ml of the stock solution (2) into a 100-ml volumetric flask and make up to the mark.

Remark:
- Frequently, Pb determinations go along with Cd determinations. If so, a mixed standard solution may be prepared with 100 mg/l Pb and 10 mg/l Cd (see 5.2.1).

Standard series

Pipette 0–0.50–1.00–2.00–3.00–4.00 ml of the standard solution (3) into 100-ml volumetric flasks. Add about 70 ml water, and 10 ml of 1.2 M hydrochloric acid (1). Make up to volume with water. This standard series has Pb concentrations of 0–0.5–1–2–3–4 mg/l.

Procedure

Nebulize the sample digests, the blank digests and the standard series into a just blue (stoichiometric) air-acetylene flame and measure the absorbance at 217.0 nm. Plot a calibration curve and read the lead concentrations.

Remarks:
- The calibration curve is slightly bent towards the x-axis, therefore the curve should not be generated by means of linear regression.
- Background correction should be applied.
- Lead absorbance can also be measured at a wavelength of 283.3 nm. In that case, the sensitivity is only half of that at 217.0 nm, whereas the noise and the background do not differ appreciably.

Calculation

The lead content of the dried plant material, expressed in mg/kg Pb, is calculated by

$$(a - b) \times \frac{V}{W}$$

in which
a = concentration of lead in the sample digest, in mg/l;
b = concentration of lead in the blank digest, in mg/l;

V = total volume of digest at the end of the digestion procedure, in ml;
w = weight of plant material sample, in g.

REFERENCES

J.J. van der Lee, E. Temminghoff, V.J.G. Houba and I. Novozamsky. Background corrections in the determination of Cd and Pb by flame AAS in plant and soil samples with high Fe levels. Appl. Spectrosc. *41* (1987) 388–390.

5.8.2 DETERMINATION OF **LEAD** BY FURNACE AAS

Field of application

This determination is intended to be carried out on digest 5.0 (dry-ashing followed by HF treatment).

It is intended for samples that are very low in Pb (< 25 mg/kg).

Principle

The sample is brought into a graphite furnace, where it is subsequently dried, ashed and vaporized by electrical heating. The lead atoms thus formed absorb radiation from a hollow-cathode lamp. The absorbance is measured at a wavelength of 283.3 nm while applying Zeeman background correction.

Remarks:

- When using Zeeman background correction, no extraction with an organic sequestering agent into xylene is needed to eliminate matrix interferences.
- Whereas the flame AAS determination (5.8.1) uses 217.0 nm as wavelength of measurement, the graphite furnace determination has a better signal-to-noise ratio at 283.3 nm and less background.

Apparatus

- Atomic absorption spectrophotometer, with electrothermal atomization device (graphite furnace) and with a device for correcting or measuring background absorption, preferably a Zeeman background correction system.

Reagents

(1) Hydrochloric acid, c(HCl) = 1.2 mol/l. Add 100 ml of concentrated hydrochloric acid (36%) to about 400 ml water and make up to 1 litre.
(2) Hydrochloric acid, c(HCl) = 0.12 mol/l. Dilute the 1.2 M hydrochloric acid (1) 1+9 (v/v).

(3) 1-Butanol (= n-butylalcohol).

(4) Stock solution, Pb concentration 1000 mg/l. Dissolve 1.5985 g of lead nitrate, $Pb(NO_3)_2$, standardized according to Appendix A-23, in some water in a 1000-ml volumetric flask. Add 10 ml of concentrated nitric acid (65%) and make up to the mark.

(5) Standard solution, Pb concentration 1 mg/l. Pipette 1.00 ml of the stock solution (4) into a 1000-ml volumetric flask. Add 1 ml of concentrated nitric acid (65%) and make up to the mark with water.

Remark:

- Frequently, Pb determinations go along with Cd determinations. If so, a mixed diluted standard solution may be prepared with 1000 μg/l Pb and 50 μg/l Cd in 0.12 M hydrochloric acid.

Standard series

Pipette 0–0.50–1.00–1.50–2.00 ml of the standard solution (5) into 100-ml erlenmeyer flasks. If Cd is also to be determined, pipette 0–0.50–1.00–1.50–2.00 ml of a 50 μg/l Cd solution (see 5.2.2) into the same erlenmeyer flasks. Dilute to about 30 ml with 0.12 M hydrochloric acid (2); mix immediately after addition of acid.

Procedure

Transfer (part of) the digests and the standard series to polythene cups that fit in the automatic sampler of the atomic absorption spectrophotometer. Add 0.20 ml of 1-butanol (3) and mix thoroughly with an electric mini-stirrer. Heat the extract in a graphite furnace according to an appropriate time-temperature programme (see remark). Measure the absorbance at 283.3 nm in the atomization phase, in the cleaning phase, and at room temperature at the end of the cycle; use Zeeman background correction. Plot a calibration curve and read the lead concentrations.

Remarks:

- When a sample gives a higher absorbance than the highest standard, it should be diluted with the zero standard solution (dilution factor = f); this can be programmed in the automatic sampler.
- The calibration curve is slightly bending towards the x-axis.
- Samples with high Fe content may give too low results when applying a deuterium background correction system.
- The absorbance reading in the cleaning phase is meant as a check on memory effects. If these are observed, the results of the last and the second last sample should be discarded. With the programme used in our laboratory (see remark), memory effects are only observed if something is out of order.
- After completion of each firing, the absorbance is read at room temperature as an indication of baseline drift. This value is used as a correction on the reading in the atomization phase.
- Every sample should be measured at least three times, since the precision of this technique is low.
- The wash solution of the automatic sampler contains 2-propanol (2% in water) in order to lower its surface tension.

- The measurements can be performed with any ETA-AAS system. The present method was worked out using a Varian SpectrAA 300 system. The optimum programme – after 30 min warming up – is as follows:

phase	I (dry)	II (char)	III (atomize)	IV (clean)	"V" (base line)
temperature (°C)	160	500	2300	2500	ambient
rate	1	0	0	0	—
hold time (s)	30	45	10	3	—
gas stop (s)	—	—	10	—	—
read	—	—	yes	yes	yes

- When using other instruments, the optimum temperature values to be set may differ from the values given above, because the temperatures mentioned here are instrument settings instead of real temperature values.

Calculation

The lead content of the dried plant material, expressed in mg/kg Pb, is calculated by

$$\frac{(a - b) \times f \times V}{w}$$

in which
a = concentration of lead in the sample digest, in µg/l;
b = concentration of lead in the blank digest, in µg/l;
f = dilution factor for the measuring solution;
V = total volume of digest at the end of the digestion procedure, in ml;
w = weight of plant material sample, in g.

5.8.3 DETERMINATION OF **LEAD** BY FURNACE AAS AFTER LIQUID-LIQUID EXTRACTION

Field of application

This determination is intended to be carried out on digest 5.0 (dry-ashing followed by HF treatment).

It is intended for samples that are very low in Pb (< 25 mg/kg).

Method 5.8.3 should only be applied when no Zeeman background correction system is available.

Principle

Lead ions are sequestered by a carbamate, and the sample is extracted into xylene. The extract is subsequently dried, ashed and vaporized by electrical heating in a graphite furnace. The lead atoms thus formed absorb radiation from a hollow-cathode lamp. The absorbance is measured at a wavelength of 283.3 nm.

Remarks:

- The extraction serves two purposes at a time: a large part of the matrix interferences is eliminated, and the lead is more concentrated.
- Since not all interferences will have disappeared, background correction is necessary.
- Whereas the flame AAS determination (5.8.1) uses 217.0 nm as wavelength of measurement, the graphite furnace determination has a better signal-to-noise ratio at 283.3 nm and less background.

Apparatus

- Erlenmeyer flasks with ground glass necks and PTFE stoppers.
- Linear shaker (see Fig. 5).
- Water repellent filter paper.
- Polythene cups.
- Atomic absorption spectrophotometer, with electrothermal atomization device (graphite furnace) and with a device for correcting or measuring background absorption.

Reagents

(1) Hydrochloric acid, c(HCl) = 1.2 mol/l. Add 100 ml of concentrated hydrochloric acid (36%) to about 400 ml water and make up to 1 litre.
(2) Hydrochloric acid, c(HCl) = 0.12 mol/l. Dilute the 1.2 M hydrochloric acid (3) 1 + 9 (v/v).
(3) DDDTC solution. Dissolve 5 g of diethylammonium-N,N-diethyldithiocarbamate, $C_9H_{22}N_2S_2$, in 100 ml xylene. Stir with a magnetic stirrer until all crystals are dissolved; this will take about 45 minutes. The solution can be kept overnight at room temperature.
(4) 1-Butanol (= n-butylalcohol).
(5) Stock solution, Pb concentration 1000 mg/l. Dissolve 1.5985 g of lead nitrate, $Pb(NO_3)_2$, standardized according to Appendix A-23, in some water in a 1000-ml volumetric flask. Add 10 ml of concentrated nitric acid (65%) and make up to the mark.
(6) Standard solution, Pb concentration 1 mg/l. Pipette 1.00 ml of the stock solution (4) into a 1000-ml volumetric flask. Add 1 ml of concentrated nitric acid (65%) and make up to the mark with water.

Remarks:

- Frequently, Pb determinations go along with Cd determinations. If so, a mixed diluted standard solution may be prepared with 1000 µg/l Pb and 50 µg/l Cd in 0.12 M hydrochloric acid.
- The solid DDDTC should be stored in a refrigerator in order to prevent deterioration.

Standard series

Pipette 0–0.50–1.00–1.50–2.00 ml of the standard solution (6) into 100-ml erlenmeyer flasks. If Cd is also to be determined, pipette 0–0.50–1.00–1.50–2.00 ml of a 50 μg/l Cd solution (see 5.2.3) into the same erlenmeyer flasks. Dilute to about 30 ml with 0.12 M hydrochloric acid (2); mix immediately after addition of acid.

Procedure

(a) Pipetting of digests
Pipette D ml (D varying between 1 and 30 ml, according to the expected lead concentrations) of the sample digests into 100-ml erlenmeyer flasks; pipette also E ml (E = highest D) of the blank digests into 100-ml erlenmeyer flasks. Dilute to 30 ml with 0.12 M hydrochloric acid (2).

(b) Extraction
Add both to standard series, sample digests and blank digests 10.00 ml of the DDDTC solution (3) and close the erlenmeyer flasks with a PTFE stopper. Shake mechanically (see Fig. 5) during 4 min (stroke length 9 cm, 150 oscillations/min). Suck off the water phase (see Fig. 6) to such an extent that the remaining water will not occupy more than half of the filter area, and filter over water repellent filter paper.

Shake an extra amount of five portions of the zero standard for future dilution purposes; mix the five extracts to make a uniform zero standard (pool).

After the extraction, the standard series has lead concentrations of 0–50–100–150–200 μg/l. (If cadmium was also put in, its concentrations are 0–2.5–5–7.5–10 μg/l).

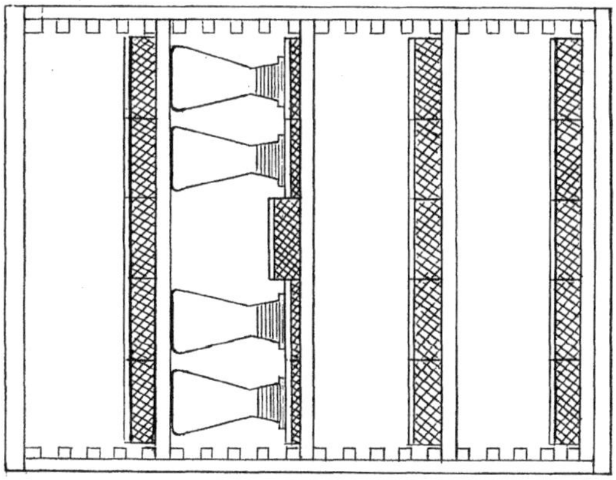

Fig. 5. Rack & erlenmeyer flasks (top view).

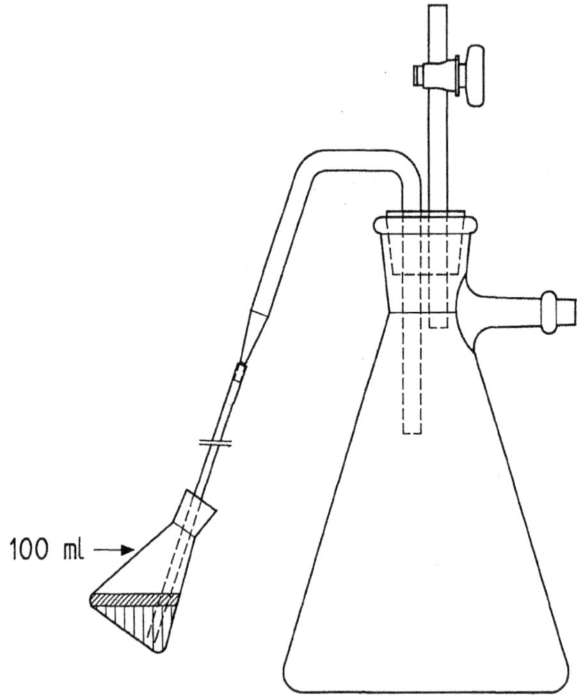

Fig. 6. Sucking off the (lower) water phase.

(c) Measurement

Pipette 1.00 ml of the extracts obtained at (b) into polythene cups that fit in the automatic sampler of the atomic absorption spectrophotometer. Add 0.20 ml of 1-butanol (4) and mix thoroughly with an electric mini-stirrer.

Heat the extract in a graphite furnace according to an appropriate time-temperature programme (see remark). Measure the absorbance at 283.3 nm in the atomization phase, in the cleaning phase, and at room temperature at the end of the cycle; use background correction.

Plot a calibration curve and read the lead concentrations.

Remarks:

- Samples and standard series should be shaken at one time, or at least directly after each other, with the same shaking machine.
- When a sample gives a higher absorbance than the highest standard, it should be diluted with the extract of the zero standard solution (dilution factor = f). If a more than tenfold dilution would be required, then the extraction should be repeated with less digest pipetted (D ml).
- The calibration curve is slightly bending towards the x-axis.
- Xylene vapours are rather poisonous. Work in a fume hood.
- For samples with high Fe content, an appreciable amount of Fe can be transferred into the organic phase. This may give too low results when applying a deuterium background correction system.
- The absorbance reading in the cleaning phase is meant as a check on memory effects. If these are observed, the results of the last and the second last sample should be discarded.

With the programme used in our laboratory (see remark), memory effects are only observed if something is out of order.

- After completion of each firing, the absorbance is read at room temperature as an indication of baseline drift. This value is used as a correction on the reading in the atomization phase.
- Every sample should be measured at least three times, since the precision of this technique is low.
- Clean all glassware, polythene cups and PTFE stoppers as follows: After use, wash once with ethanol 96% to remove the remaining xylene. Let drain, rinse with water and shake with 4 M HNO_3 for 5 minutes (or leave overnight). Then rinse with water and twice with ethanol 96%. Allow to dry by leaving at room temperature in an inverted position.
- The wash solution of the automatic sampler contains 2-propanol (2% in water) in order to lower its surface tension.
- The complex of Pb-DDDTC in xylene is stable for about 24 hours, and can therefore be measured on "extraction day" or the day after.
- The measurements can be performed with any ETA-AAS system. The present method was worked out using a Perkin-Elmer system, consisting of a model 460 AA spectrophotometer, a HGA 76 B graphite furnace, an AS-1 automatic injection unit annex sampler and a PRS-10 printer. With a lamp current of 5 mA, a gain of 40, an integration time of 10 s, background correction on (mode AA minus BG), slit 0.7 nm Alt and peak height mode, the optimum programme – after 30 min warming up – is as follows:

phase	I (dry)	II (char)	III (atomize)	IV (clean)	"V" (base line)
temperature (°C)	160	500	2300	2500	ambient
rate	1	0	0	0	—
hold time (s)	30	45	10	3	—
gas stop (s)	—	—	10	—	—
read	—	—	yes	yes	yes

- When using other instruments, the optimum temperature values to be set may differ from the values given above, because the temperatures mentioned here are instrument settings instead of real temperature values.

Calculation

The lead content of the dried plant material, expressed in µg/kg Pb, is calculated by

$$\left\{\frac{E \times a}{D} - b\right\} \times \frac{10 \times f \times V}{E \times w}$$

in which

a = concentration of lead in the xylene extract of the sample digest, in µg/l;
b = concentration of lead in the xylene extract of the blank digest, in µg/l;
f = dilution factor for the measuring solution (remark);
V = total volume of digest at the end of the digestion procedure, in ml;
D = volume of sample digest pipetted for extraction, in ml;
E = volume of blank digest pipetted for extraction, in ml; (E = highest D)
w = weight of plant material sample, in g.

Remark:

- If f = 1, D = 30, V = 100 and w = 2, then the lead content is (a — b)/60.

5.9 DETERMINATION OF **ZINC** BY FLAME AAS

Field of application

This determination is intended to be carried out on digest 5.0 (dry-ashing followed by HF treatment).

The zinc content of the dried plant material should be higher than 4.5 mg/kg.

Principle

The sample is nebulized into an air-acetylene flame, where it is vaporized; zinc compounds are atomized and the zinc atoms thus formed absorb radiation from a hollow-cathode lamp. The absorbance is measured at a wavelength of 213.9 nm.

Apparatus

- Atomic absorption spectrophotometer.

Reagents

(1) Hydrochloric acid, c(HCl) = 0.12 mol/l. Add 10 ml of concentrated hydrochloric acid (36%) to about 400 ml water and make up to 1 litre.
(2) Stock solution, Zn concentration 400 mg/l. Dissolve 1.759 g of zinc sulphate heptahydrate, $ZnSO_4 \cdot 7H_2O$ (standardized according to Appendix A-24) in about 600 ml water in a 1000-ml volumetric flask. Add 10 ml of concentrated hydrochloric acid (36%) and make up to volume.
(3) Standard solution, Zn concentration 20 mg/l. Pipette 25.0 ml of the stock solution (2) into a 500-ml volumetric flask and make up to volume with 0.12 M hydrochloric acid (1).

Remark:
- Frequently, Zn determinations go along with Mn determinations. If so, a mixed standard solution may be prepared with 20 mg/l Zn and 50 mg/l Mn (see 5.6.1).

Standard series

Pipette 0–1.00–2.00–4.00–6.00–8.00–10.00 ml of the standard solution (3) into 100-ml volumetric flasks and make up to volume with 0.12 M hydrochloric acid (1). This standard series has Zn concentrations of 0–0.2–0.4–0.8–1.2–1.6–2.0 mg/l.

Procedure

Nebulize the sample digests, the blank digests and the standard series into a just blue (stoichiometric) air-acetylene flame and measure the absorbance at 213.9 nm. Plot a calibration curve and read the zinc concentrations.

Remarks:

- The calibration curve is slightly bent towards the x-axis, so that calculation by means of lineair regression is not allowed.
- Although the chosen wavelength is short, no background correction has been found necessary when using 0.15 M perchloric acid. For other media, background correction may be required.
- No interferences were observed after spiking the digests with several salts, so as to reach final concentrations of 100 mg/l Mg, 30 mg/l Fe, 20 mg/l Al + 400 mg/l PO_4, or 30 mg/l Fe + 20 mg/l Al + 400 mg/l PO_4.
- Do not use rubber stoppers, since these will release zinc.

Calculation

The zinc content of the dried plant material, expressed in mg/kg Zn, is calculated by

$$(a - b) \times \frac{V}{w}$$

in which

a = concentration of zinc in the sample digest, in mg/l;
b = concentration of zinc in the blank digest, in mg/l;
V = total volume of digest at the end of the digestion procedure, in ml;
w = weight of plant material sample, in g.

6 Digestion by **dry-ashing** in the presence of CaO and determination of B

6.0 DIGESTION BY **DRY-ASHING** IN THE PRESENCE OF CAO

Field of application

This digestion is intended solely for the colorimetric determination of B.

Principle

The organic plant material matrix is destroyed by controlled heating. Boron compounds are kept by calcium oxide.

Apparatus

- Porcelain crucibles.
- Hot plate.
- Muffle furnace.
- Polythene stirring rods.

Reagents

(1) Calcium oxide, solid, finely powdered.
(2) Sulphuric acid, $c(H_2SO_4) = 0.5$ mol/l. Add carefully, while swirling, 28 ml of concentrated sulphuric acid (96%) to about 400 ml water. Allow to cool and make up to 1 litre.

Procedure

Weigh, to the nearest 0.001 g, approximately 1 g of the dried plant material sample in a porcelain crucible. Add 100 mg of calcium oxide (1) and mix thoroughly until no CaO particles can be distinguished any more from the plant material.

Place the crucible on a preheated hot plate (about 200 °C) and heat until the sample has charred completely. Take care that the sample does not start burning. Cool the charred sample and place it in a muffle furnace. Raise the temperature to 500 °C and maintain this temperature for 1.5 h. Take the crucible out of the furnace and allow to cool.

Add 10 ml of sulphuric acid (2). Let stand for 30 min and stir several times during this period. Filter over coarse paper into a polythene tube.

Remarks:

- Part of the sulphuric acid is consumed by the calcium oxide, so that finally the net acid strength is approximately 0.35 M.
- With the precipitate of $CaSO_4$, that has been formed after the addition of sulphuric acid, other compounds will coprecipitate so that interferences are diminished.

REFERENCES

U.C. Gupta. The boron determination of some plant materials as determined with and without adding CaO before ashing. Plant Soil *26* (1967) 202–204.

6.1 DETERMINATION OF **BORON** BY COLORIMETRY

Field of application

This determination is intended to be carried out on digest 6.0 (dry-ashing with CaO).

The boron content of the dried plant material should be higher than 1 mg/kg.

For samples with high Fe levels method 6.2 is recommended.

Principle

Borate ions form a red coloured compound with Azomethine-H at pH 4–5. The absorbance of the complex is measured at a wavelength of 430 nm.

Remarks:

- Interference from Cu and Al is prevented by addition of EDTA.
- Thioglycollic acid is used to mask Fe; however, a slight Fe interference may remain because most Fe complexes are coloured.

Apparatus

- Polythene test tubes with covers (because of bad smell).
- Colorimeter with interference filter.

Reagents

(1) Sulphuric acid, $c(H_2SO_4)$ = 1.0 mol/l. Add carefully, while swirling, 56 ml of concentrated sulphuric acid (96%) to about 400 ml water. Allow to cool and make up to 1 litre.
(2) Thioglycollic acid, 98% (mercaptoacetic acid, $C_2H_4O_2S$, d = 1.32 g/ml).
(3) Azomethine-H (4-hydroxy-5[salicylidene-amino]-2,7-naphtalene disulphonic acid, $C_{17}H_{13}NO_8S_2$). This chemical can be purchased commercially, but may be prepared readily in the laboratory (see App. A-25).
(4) Azomethine-H mixture. Dissolve 2 g of ascorbic acid and 0.9 g of Azomethine-H (3) in some water and make up to 100 ml in a polythene flask. Prepare fresh daily.
(5) Mixed buffer solution, pH 4.9. Dissolve 100 g of ammonium acetate, CH_3COONH_4, in 160 ml water. Add 50 ml of concentrated acetic acid (100%) and mix. Dissolve 2.68 g of $Na_2EDTA \cdot 2H_2O$ in the mixture. Add 2.4 ml of thioglycollic acid (2), mix well and allow to stand overnight. Store in a polythene bottle. Prepare fresh weekly.
(6) Stock solution, B concentration 100 mg/l. Dissolve 0.8819 g of borax (sodium tetraborate decahydrate, $Na_2B_4O_7 \cdot 10H_2O$)(see remark) in some water in a 1000-ml volumetric flask and make up to volume. Store in a polythene bottle.
(7) Standard solution, B concentration 5 mg/l. Dilute 25.0 ml of the stock solution (6) with water to 500 ml. Store in a polythene bottle.

Remarks:
- Borax easily loses crystal water on standing. Recrystallize it to be sure that the salt contains 10 molecules of crystal water. Procedure: See App. A-26.
- To avoid any problems with borax, the use of boric acid, H_3BO_3, may be considered. This compound should be used as such, because drying at 105 °C, or even storage in a desiccator, causes loss of water and conversion to metaboric acid, HBO_2. A stock solution of 100 mg/l B will contain 0.5719 g H_3BO_3 per litre.
- To prevent desorption of borate from the glassware to be used, this must be filled with 4 M HNO_3 and left overnight before cleaning in the usual manner. For the same reason, reagents and standard solutions should be transferred to polythene vessels directly after preparation.
- The mixed buffer solution (5) should be used only after standing overnight, since the colour to be measured fades when using the freshly prepared buffer. It is assumed that the thioglycollic acid (2) is responsible for this effect.

Standard series

Pipette 0–5.00–10.00–20.00–30.00–40.00 ml of the standard solution (7) into 100-ml volumetric flasks and add 35 ml of 1.0 M sulphuric acid (1). Make up to volume with water. Store in polythene bottles. This standard series has B concentrations of 0–0.25–0.5–1–1.5–2 mg/l.

Procedure

Pipette 2.00 ml of the sample digests, the blank digests and the standard series into polythene test tubes. Add 4.0 ml of mixed buffer solution (5) and mix thoroughly. Add 2.0 ml of Azomethine-H mixture (4) and mix again thoroughly. Measure the absorbance in a 1-cm cuvette at a wavelength of 430 nm after 30 min and within 1.5 h. Plot a calibration curve and read the boron concentrations.

Remarks:

- The calibration curve should be linear. The zero standard normally reads 0.190–0.225 Abs and the highest standard usually reads 0.550–0.590 Abs.
- Instead of the 410 nm mentioned in literature, 430 nm was chosen for absorbance measurement because the background absorption of the reagent is much lower there.
- No interferences were found with at least 1000 mg/kg Fe, 1000 mg/kg Al and 50 mg/kg Cu (levels in dried plant material).
- Laboratory coats fresh from the laundry might release boron (from the perborate used in washing powder).

Calculation

The boron content of the dried plant material, expressed in mg/kg B, is calculated by

$$(a - b) \times \frac{V}{w}$$

in which

a = concentration of boron in the sample digest, in mg/l;
b = concentration of boron in the blank digest, in mg/l;
V = total volume of digest at the end of the digestion procedure, in ml;
w = weight of plant material sample, in g.

REFERENCES

B. Wolf. The determination of boron in soil extracts, plant materials, composts, manures, water and nutrient solutions. Commun. Soil Sci. Plant Anal. *2* (1971) 363–374.

S.S. de Bes, P.A. den Dekker and P.A. van Dijk. The determination of boron with Azomethine-H in plant material, soil and water. Internal communication of the Glasshouse Crops Research Station, Naaldwijk, The Netherlands (1973) (in Dutch).

B.Wolf. Improvement of the Azomethine-H method for the determination of boron. Commun. Soil Sci. Plant Anal. *5* (1974) 39–44.

6.2 DETERMINATION OF **BORON** BY COLORIMETRY AFTER LIQUID-LIQUID EXTRACTION

Field of application

This determination may be carried out on digest 6.0 (dry-ashing with CaO).
The method is only recommended in case of high Fe levels because method 6.1 is normally sensitive enough and anyhow more rapid than method 6.2.

Principle

The complex formed by borate and 2-ethyl-1,3-hexanediol is extracted into xylene at pH < 1. The borate is back-extracted into alkaline aqueous solution and then determined through formation of the red Azomethine-H complex, of which the absorbance is measured at a wavelength of 430 nm.

Remark:

- By the liquid-liquid extraction the interferences from Al, Cu and Fe (see determination 6.1) have been eliminated, so that neither EDTA nor thioglycollic acid need to be used. As another advantage, the colour is not affected during a relatively long period.

Apparatus

- Linear shaker.
- Separating funnels with PTFE cocks and PTFE stoppers (see Fig. 1).

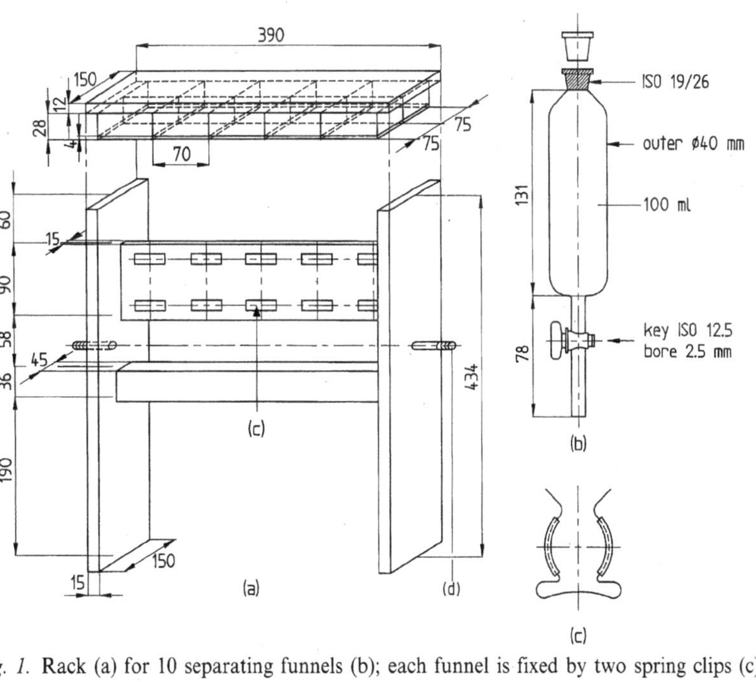

Fig. 1. Rack (a) for 10 separating funnels (b); each funnel is fixed by two spring clips (c).

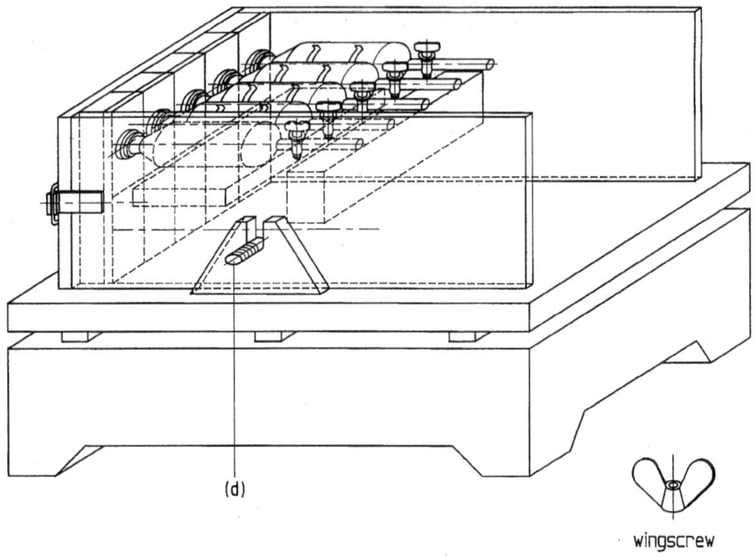

Fig. 2. Rack with separating funnels placed in the shaker; the rack is fastened with wingscrews at position (d).

- Rack for separating funnels to be placed in the shaker (see Fig. 2).
- Polythene test tubes with covers (because of bad smell).
- Colorimeter with interference filter.

Reagents

(1) Sulphuric acid, c(H_2SO_4) = 0.7 mol/l. Add carefully, while swirling, 40 ml of concentrated sulphuric acid (96%) to about 400 ml water. Allow to cool and make up to 1 litre.
(2) Sulphuric acid, c(H_2SO_4) = 0.07 mol/l. Dilute 100 ml of 0.7 M sulphuric acid (1) to 1 litre.
(3) Sodium hydroxide solution, c(NaOH) = 0.1 mol/l. Dissolve 4 g of sodium hydroxide, NaOH, in 1 litre water.
(4) EHD solution, 5%. Dissolve 25 ml of 2-ethyl-1,3-hexanediol, $C_8H_{18}O_2$, in 500 ml xylene.
(5) Azomethine-H (4-hydroxy-5[salicylidene-amino]-2,7-naphtalene disulphonic acid, $C_{17}H_{13}NO_8S_2$). This chemical can be purchased commercially, but may be prepared readily in the laboratory (see App. A-25).
(6) Azomethine-H mixture. Dissolve 2 g of ascorbic acid and 0.9 g of Azomethine-H (5) in some water and make up to 100 ml in a polythene flask. Prepare fresh daily.
(7) Buffer solution, pH 4.9. Transfer 118 g of ammonium acetate, CH_3COONH_4, with water into a 250-ml volumetric flask. Add 59 ml of concentrated acetic acid (100%), dissolve and make up to volume. Store in a polythene bottle.

(8) Stock solution, B concentration 100 mg/l. Dissolve 0.8819 g of borax (sodium tetraborate decahydrate, $Na_2B_4O_7.10H_2O$)(see remark) in some water in a 1000-ml volumetric flask and make up to volume. Store in a polythene bottle.

(9) Standard solution, B concentration 5 mg/l. Dilute 25.0 ml of the stock solution (8) with water to 500 ml. Store in a polythene bottle.

Remarks:

- Borax easily loses crystal water on standing. Recrystallize it to be sure that the salt contains 10 molecules of crystal water. Procedure: See App. A-26.
- To avoid any problems with borax, the use of boric acid, H_3BO_3, may be considered. This compound should be used as such, because drying at 105 °C, or even storage in a desiccator, causes loss of water and conversion to metaboric acid, HBO_2. A stock solution of 100 mg/l B will contain 0.5719 g H_3BO_3 per litre.
- To prevent borate contamination from the glassware to be used, reagents and standard solutions should be transferred to polythene vessels directly after preparation.

Standard series

Pipette 0–0.50–1.00–2.00–3.00–4.00 ml of the standard solution (9) into 100-ml separating funnels (Fig. 1) and follow the procedure. This standard series has B concentrations of 0–0.25–0.5–1–1.5–2 mg/l in the NaOH phase.

Procedure

(a) Extraction

Place the 100-ml separating funnels in the rack (see Fig. 1) and adjust the height so that later on, while shaking, the stoppered funnels will be pushed against the cover in order to prevent leakage. **Do not remove the funnels any more.**

Pipette 5.00 ml of digest 6.0 into the separating funnels. Add 2.0 ml of 0.7 M sulphuric acid (1) and water to a total volume of 27 ml. Prepare also the standard series and a blank digest.

Add 30 ml of EHD solution (4); stopper the funnels, fasten the cover and mount the rack horizontally on the shaker (see Fig. 2). Shake during 5 min with 150 oscillations per minute and stroke length 9 cm. Remove the rack from the shaker, place it upright and wait until the layers have separated.

Remove the stopper and apply gentle suction at the neck of the separating funnel. Then open the stopcock carefully so that some air enters the funnel through the stem and any drops of xylene trapped in the stem are bubbled up to join the bulk of the organic phase.

Discard the water layer and wash two times carefully with 10-ml portions of 0.07 M sulphuric acid (2).

(b) Back-extraction

Add 10 ml of sodium hydroxide solution (3) and shake again during 5 min. Wait until the phases have separated. Remove again any xylene drops from the stem in the way described above. Collect the aqueous phase in a polythene bottle.

(c) Measurement

Pipette 2.00 ml of the alkaline aqueous phase into a polythene test tube. Add subsequently 1.0 ml of 0.7 M sulphuric acid (1), 3.0 ml of buffer solution (7) and 2.0 ml of Azomethine-H mixture (6); mix thoroughly after each addition. Wait for 30 minutes. Measure the absorbance in a 1-cm cuvette at a wavelength of 430 nm. Plot a calibration curve and read the boron concentrations.

Remarks:

- The rack has been designed such that the separating funnels never need to be removed. Thus, the rubber tubing around the spring clips is intended to fix the funnels at the right position, and the height of the rack is such that there is enough free space below the funnels to place volumetric flasks there.
- Samples and standard series should be shaken together at the same time, on a powerful linear shaker at the prescribed conditions. Otherwise, the extraction yield may be variable and low. If the conditions of 9 cm stroke length and/or 150 strokes per minute cannot be fulfilled, the time of shaking should be prolonged; the optimum time should be established by shaking standard series for different periods.
- The calibration curve should be linear.
- To prevent desorption of borate from the glassware to be used, this must be filled with 4 M HNO_3 and left overnight before cleaning in the usual manner.
- Instead of the 410 nm mentioned in literature, 430 nm was chosen for absorbance measurement because the background absorption of the reagent is much lower there.
- Xylene and its vapours are rather poisonous. Work in a fume hood.
- Laboratory coats fresh from the laundry might release boron (from the perborate used in washing powder).

Calculation

The boron content of the dried plant material, expressed in mg/kg B, is calculated by

$$(a - b) \times \frac{10}{D} \times \frac{V}{w}$$

in which

a = concentration of boron in the sample digest, in mg/l;
b = concentration of boron in the blank digest, in mg/l;
D = volume of digest taken for extraction, in ml;
V = total volume of digest at the end of the digestion procedure, in ml;
w = weight of plant material sample, in g.

REFERENCES

B. Wolf. The determination of boron in soil extracts, plant materials, composts, manures, water and nutrient solutions. Commun. Soil Sci. Plant Anal. *2* (1971) 363–374.

S.S. de Bes, P.A. den Dekker and P.A. van Dijk. The determination of boron with Azomethine-H in plant material, soil and water. Internal communication of the Glasshouse Crops Research Station, Naaldwijk, The Netherlands (1973) (in Dutch).
B. Wolf. Improvement of the Azomethine-H method for the determination of boron. Commun. Soil Sci. Plant Anal. *5* (1974) 39–44.

7 Extraction with **water** and determination of Cl, NO_3, NO_2, SO_4

7.0 EXTRACTION WITH **WATER**

Field of application

This extraction is meant for the determination of Cl^-, NO_2^-, NO_3^- and SO_4^{2-}.

Principle

The anions that are "free" in the plant material are extracted as such by water.

Apparatus

- Metal weighing funnel (see Fig.1)
- Shaker, linear or end-over-end.

Remark:

- Dried plant material may easily stick to glass when the relative humidity of the air is low. These home-made weighing funnels (stainless steel or aluminium) do not show this effect.

Reagents

(1) Water.
(2) Fluted filter paper, fine, diameter 15 cm.

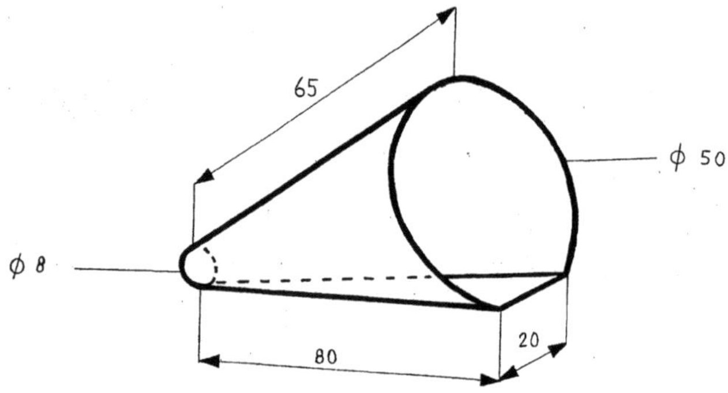

Fig. 1. Weighing funnel.

Procedure

Weigh, to the nearest 0.001 g, approximately 0.5 g of the dried plant material sample in a metal weighing funnel and transfer the sample to a 50-ml erlenmeyer flask. Add 25 ml water and shake for 30 min at ambient temperature. Filter over fine paper. Collect the filtrate and filter this very portion of filtrate again over *the same* piece of paper. Use the last filtrate for analysis.

Remarks:

- The first filtration will clog the paper just enough to obtain a more clear filtrate at the second time.
- The determinations must be carried out on the same day as the extraction, since the filtrates will rapidly turn turbid due to mould growth.
- If all three anions mentioned are to be determined, the extraction has to be done in duplicate or triplicate and the extracts combined in order to obtain a sufficient amount of test sample.
- See determination 7.1.2 if only chloride has to be measured.
- See determination 8.1 if only nitrate has to be measured.

7.1.1 DETERMINATION OF **CHLORIDE** BY COULOMETRIC TITRATION

Field of application

This determination may be carried out on extract 7.0 (water).

Principle

Chloride is titrated with silver ions, which are generated from a silver anode at constant current. The end point is detected by a sudden increase in current through a separate set of silver electrodes, caused by the first excess of free silver ions; this current activates a trigger circuit to stop the pulse counter. The total number of pulses counted is a direct measure of the amount of chloride in the sample.

Remarks:

- In a coulometric titration at constant current, the total quantity of electricity passed is derived from the product current × time. In the present procedure, the apparatus used is calibrated directly in millimoles of chloride.
- The solution to be measured is acidified to allow a smooth reaction at the reference electrode, and also to prevent adsorption of silver ions on the precipitate and to prevent the precipitation of other silver salts, like phosphate and carbonate.
- Gelatin is added to the solution to keep the precipitate well suspended. In that way, all newly generated silver ions come instantaneously into contact with the precipitate (see remark).

Apparatus

- Coulometer, fitted for titrations at constant current.
- Working electrode, silver.
- Reference electrode, silver.
- Detection electrodes, silver.

Remark:

- In the authors' laboratory, a so-called "Chlor-O-Counter" is used; this is a Dutch-made coulometer which is especially adapted for rapid chloride determinations. The procedure is written for the use of this particular instrument, but in principle any commercially available coulometer will serve the purpose.

Reagents

(1) Acid mixture. Mix 8.0 ml of concentrated nitric acid (65%) with 100 ml of concentrated acetic acid (100%) and dilute to 1000 ml with water.

(2) Gelatin solution. Weigh out 600 mg of gelatin (granules of purity grade "album") in a 100-ml beaker. Add 50 ml water and soak for 2 h at room temperature. Thereafter, heat on a water bath or in a beaker with hot water until the gelatin has dissolved. Add 10 mg of thymol (to prevent mould growth) and 10 mg of thymol blue; make up to 100 ml. Store in a refrigerator but do not freeze. Prepare fresh weekly.

(3) Stock solution, Cl concentration 100 mmol/l. Dissolve 5.844 g of sodium chloride, NaCl, (pretreated according to App. A-4) in water in a volumetric flask of 1000 ml and make up to the mark.

(4) Standard solution, Cl concentration 10 mmol/l. Pipette 10.0 ml of the stock solution (3) into a 100-ml volumetric flask and dilute to volume.

Standard series

No standard series is needed, since the number of pulses (representing the total amount of electricity consumed) is directly proportional to the amount of chloride. Only one standard solution must be titrated to establish the actual relation.

Procedure

a) Pretreatment of the titration medium

Check whether both electrode couples are scrupulously clean, in particular the detection electrodes; if necessary, cleanse them with a cleansing powder and wipe with tissue paper. Check also whether the generator electrode is still thick enough (at least 2 mm).

Bring into a tall-shaped 50-ml beaker:

- 20 ml of acid mixture (1)
- 1 ml of gelatin solution (2)
- 2 ml of standard solution (4).

Place the beaker on the titration stand, and immerse the electrodes and the stirrer. Adjust the current at the highest possible value (here: 80×10^{-9} mol Cl^- per pulse, corresponding with 96 mA), and press the pushbutton "mixing". After a few seconds, press the key "titrating". The titration process can be followed by observing the pulse counter and the formation of precipitate. When the titration is finished, the counter stops automatically. By pressing the key "pipetting", the stirring is stopped and the counter is put to zero again; the number of pulses needs not to be recorded. The titration medium is now ready for the measurements.

b) Measurement of standard solution

Pipette 1.00 ml of the standard solution (4) into the same 50-ml beaker containing the previously titrated solution. Stir for a few seconds and then start the titration with a current of 12 mA (= 10×10^{-9} mol Cl^- per pulse). When the titration is finished, record the number of pulses (about 1000) and put the counter to zero again.

Repeat this procedure two or three times and adjust the current in such a way that 1000 pulses are counted.

c) Measurement of samples

Pipette 1.00 ml of the extract or the blank into the same 50-ml beaker containing the previously titrated solution. Stir for a few seconds and then start the titration with a current of 6 mA (= 5×10^{-9} mol Cl^- per pulse). When the titration is finished, record the number of pulses. Put the counter to zero again, pipette the next sample into the same titration medium, and proceed as before. When the beaker has become completely filled, the procedure has to be started again with a fresh titration medium (see remark).

When less than 150 pulses per sample are counted, the measurement must be repeated with more sample (D ml).

Remarks:

- The silver anode slowly dissolves by forming silver chloride. The consumption of silver is, however, very small: 1 mg for 1 ml of standard solution.

- The thymol blue is red-coloured below pH 2, so that it indicates whether the titration medium is still acidic enough and also whether the gelatin solution is actually added.
- The titration medium must contain some precipitate of AgCl, so that during the sample titration the precipitate is formed rapidly enough.
- When the beaker has become completely filled after several titrations, the titration medium is normally still acidic enough. In that case, it is appropriate to empty the beaker partly and to proceed with the next samples. In this way, about twenty samples can be handled before the titration medium is "exhausted" (see remark).
- In case the precipitation starts with difficulty, or the counter begins to run irregularly, the titration medium may have become "exhausted", which means that the acid has been consumed by the reaction at the reference electrode. The procedure should then be started again with fresh titration medium.
- For chloride determinations in plant extracts, the prescribed current of 6 mA is for most samples sufficient. When high concentrations are expected, a higher current value can be chosen.

Calculation

The chloride content of the dried plant material, expressed in mmol/kg Cl, is calculated by

$$0.005 \times (a - b) \times \frac{1}{D} \times \frac{V}{w}$$

in which

a = number of pulses counted in the sample extract;
b = number of pulses counted in the blank extract;
D = volume of plant extract pipetted, in ml;
V = volume of extract used for the extraction procedure, in ml;
w = weight of plant material sample, in g.

Remarks:

- This calculation holds only for the Chlor-O-Counter in position 5 (= 5×10^{-9} mol Cl^- per pulse). If a higher current p was set, then multiply by p/5.
- In the calculation formula it is assumed that the standard solution provides exactly 1000 pulses. Otherwise, a correction factor 1000/n must be applied, where n = actual number of pulses for the standard solution.

7.1.2 DETERMINATION OF **CHLORIDE** BY POTENTIOMETRIC TITRATION

Field of application

This determination may be carried out on extract 7.0 (water).

Since the titration is not interfered by plant material particles, it may be carried out directly in the suspension; this is most appropriate when solely chloride is to be determined.

Principle

Chloride is titrated with silver nitrate in acidic medium. The end point is indicated by the sharp increase in potential of the silver wire electrode used.

Remarks:

- Nitric acid is added (i) to create better precipitation conditions for AgCl, (ii) to prevent formation of other precipitates, and (iii) to suppress any influence of differences in activity.
- The equivalence point can be determined graphically, instrumentally or by calculation. In the present procedure, a combination of the latter two was chosen: the equivalence potential is established by calculation after titrating some standard solutions, and then the samples are titrated simply up to this equivalence potential.

Apparatus

- Potentiometer (pH/mV meter), which allows electrode potential measurements between at least —500 mV and +500 mV with a precision better than 1 mV.
- Indicator electrode: silver wire (pure silver, no alloy), length 15 cm, diameter 1 mm.
- Reference electrode: Hg/Hg_2SO_4 electrode with saturated K_2SO_4 as contact electrolyte.
- Magnetic stirrer.
- Piston burette.

Remark:

- If only an analog pH/mV meter with scale divisions of 10 mV is available, it is advised to use the pH scale since this is actually an expanded mV scale.

Reagents

(1) Nitric acid, $c(HNO_3)$ = 0.30 mol/l. Dilute 21 ml of concentrated nitric acid (65%) to 1 litre with water.

(2) Silver nitrate solution, Ag^+ concentration 0.050 mol/l. Dissolve 8.50 g of silver nitrate in 1 litre water. Standardize by potentiometric titration (see "Standard series").

(3) Standard solution, Cl^- concentration 0.0500 mol/l. Dissolve 2.922 g of sodium chloride, NaCl, (pretreated according to App. A-4) in water in a volumetric flask of 1000 ml and make up to the mark.

Standard series

Notwithstanding the linear relationship, a standard series is needed to confirm whether the intercept is really zero. (The zero standard cannot be titrated directly, because there is no precipitate.) Therefore, three different volumes of the standard solution (3) are titrated (see "Procedure", b); this also serves the purpose of standardizing the silver nitrate solution in exactly the same way as the samples are measured.

Procedure

(a) Determination of the potential at the equivalence point

Pipette 1.00–2.00 5.00 ml of the standard solution (3) into 100-ml beakers. Add 40 ml of nitric acid (1). Add a magnetic bar of about 15 mm length and start stirring. Insert both electrodes; immerse the silver wire to a depth of 2 cm in the solution. Keep on stirring smoothly till the end of the titration.

Record the starting position of the burette (in ml) as well as the reading of the potentiometer (in mV). Add slowly silver nitrate solution (2) up to about 0.3 ml short of the expected equivalence point (if needed, carry out a preliminary titration). Record the position of the burette (in ml) and the potentiometer reading (in mV). Add 2 drops of silver nitrate solution and record again the volume and potential values. Continue titrating with portions of 2 drops, while recording the volume and potential values, till about 0.3 ml beyond the equivalence point.

Make a table of the values for the consumed volume of silver nitrate solution (v, in ml), the volume difference between two readings (Δv, in ml), the potentiometer reading (E, in mV), the potential difference between two readings (ΔE, in mV), the change in potential per unit change in volume of titrant (ΔE/Δv, i.e. the first derivative), and the difference between subsequent values of the first derivative (Δ2E/Δv). *It may be helpful to consult the example table at the end of this section.*

Calculate for each of the three aliquots of standard solution (3) the potential at the equivalence point according to

$$E_e = E_p + \Delta E_p \times \frac{\Delta 2E(+)}{[\Delta 2E(+)] + [\Delta 2E(-)]}$$

in which

E_e = potential at the equivalence point (assumed to coincide with the inflection point), in mV;

E_p = potential at which the last positive value of Δ2E/Δv was read, in mV;

ΔE_p = potential difference around the inflection point (= largest difference), in mV;

Δ2E(+) = last positive value of Δ2E/Δv;

Δ2E(—) = first negative value of Δ2E/Δv.

Compare the three values obtained for E_e and take the average.

(b) Standardization of silver nitrate

Calculate for each of the three aliquots of standard solution (3) the exact volume of silver nitrate solution (2) consumed up till the equivalence point by

$$v_e = v_p + \Delta v_p \times \frac{\Delta 2E(+)}{[\Delta 2E(+)] + [\Delta 2E(-)]}$$

in which

v_e = volume of titrant added at the equivalence point, in ml;

v_p = volume at which the last positive value of $\Delta 2E/\Delta v$ was read, in ml;

Δv_p = volume difference around the inflection point (i.e., the portion of titrant corresponding with ΔE_p), in ml.

For the calculation of v_e also an example is worked out at the end of this section.

Plot a calibration curve in which the pipetted volumes v_s of standard chloride solution (3) on the x-axis are plotted against the consumed volumes v_e of silver nitrate solution (2) on the y-axis. Calculate the intercept v_i (= assumed consumption of the zero standard).

Calculate the molarity of the silver nitrate solution (2) by

$$t_{AgNO3} = 0.0500 \times \frac{v_s}{(v_e - v_i)}$$

in which

t_{AgNO3} = strength of silver nitrate solution, in mol/l;

v_s = volume of standard solution pipetted, in ml.

v_e = volume of titrant added at the equivalence point, in ml;

v_i = calculated consumption of titrant for the zero standard, in ml.

(c) Measurement of chloride

Weigh, to the nearest 0.001 g, approximately 1 g of the dried plant material sample into a 100-ml beaker and add 40 ml of nitric acid (1).

Introduce a magnetic bar of about 15 mm length and start stirring. Insert both electrodes; immerse the silver wire electrode to a depth of 2 cm in the liquid. Record the starting position of the burette (a_1 ml). Add slowly silver nitrate solution (2) until the calculated value of the equivalence point (E_e mV) is reached. Record the burette position (a_2 ml) at this potential.

Remarks:

- Samples with very low chloride concentration can be recognized from their high initial potential. They should be titrated very carefully, adding the silver nitrate dropwise right from the beginning.
- If preferred, the determination may be done on two portions (≈ about 40 ml) of extract 7.0 after addition of 5 ml of concentrated nitric acid.
- Clean the silver electrode frequently; this can be done with any household silver cleaner.

Calculation

The chloride content of the dried plant material, expressed in mmol/kg Cl, is calculated by

$$1000 \times (a_2 - a_1) \times \frac{t}{w}$$

in which

$(a_2 - a_1)$ = volume of silver nitrate solution consumed, in ml;

t = concentration of silver nitrate solution, in mol/l;
w = weight of plant material sample, in g.

For further explanation, a table is given with data from a real titration of 1.00 ml of a 0.0500 M chloride solution with 0.05 M silver nitrate:

volume of titrant (v (ml))	potential reading (E (mV))	volume difference (Δv (ml))	potential difference (ΔE (mV))	first derivative (ΔE/Δv)	difference in first derivative (Δ2E/Δv)
0	— 236				
0.70	— 205				
		0.09	10	111	
0.79	— 195				+ 56
		0.09	15	167	
0.88	— 180				+ 322
		0.09	44	489	
0.97	— 136				+ 178
		0.09	60	667	
1.06	— 76				— 467
		0.09	18	200	
1.15	— 58				— 78
		0.09	11	122	
1.24	— 47				— 44
		0.09	7	78	
1.33	— 40				

For the example given in the table, the right potential values are: E_p = — 136 mV, ΔE_p = 60 mV, Δ2E(+) = + 178, and Δ2E(—) = — 467, so that the formula for E_e reads

$$E_e = -136 + 60 \times \frac{+178}{178 + 467} = -\ 120 \text{ mV}$$

In the same way, v_e can be calculated as

$$v_e = 0.97 + 0.09 \times \frac{+178}{178 + 467} = 0.995 \text{ ml}$$

7.2.1 DETERMINATION OF **NITRATE** BY DIRECT POTENTIOMETRY

Field of application

This determination is to be carried out on extract 7.0 (water).

The nitrate content of the dried plant material should be higher than 0.5 mmol/kg.

Remark:

- This method may be used for the determination of nitrate in plant material only when the nitrate content is high. At lower levels the interferences by inorganic but especially organic anions are too strong. The nitrate concentration at which the interferences are negligible depends on the plant species (see literature reference). For low concentrations of nitrate method 8.1 should be used.

Principle

After buffering the ionic strength of the water extracts by a phosphate solution, the nitrate ions are detected directly with an ion-selective electrode.

Remarks:

- Some anionic species interfere: chloride, bicarbonate, phosphate, oxalate, acetate, citrate, succinate, fumarate, malate and malonate. These interferences are partly combatted by lowering the pH to 4.0 and addition of cation exchange resin saturated with Al^{3+}; the resin lowers the pH further to about 2.5–3. Thus the ionisation is decreased and the organic anions are partly complexed.
- The addition of phosphate and Al-resin also serves the purpose of buffering the ionic strength.

Apparatus

- Potentiometer (pH/mV meter) with which a precision of better than 0.5 mV can be obtained.
- Magnetic stirrer.
- Indicator electrode: nitrate ion-selective electrode.
- Reference electrode: Hg/Hg_2SO_4 electrode with saturated K_2SO_4 as contact electrolyte.

Remark:

- In practice, a common calomel electrode can be used, provided that the chloride leakage is very small and no Ag-resin is applied.

Reagents

(1) Buffer solution, pH 4.0. Dissolve 27.2 g of potassium dihydrogen phosphate, KH_2PO_4, in some water and make up to 1 litre.
(2) Cation exchange resin. Dowex 50 W-X 8, 20–50 mesh, in H^+- or Na^+-form.
(3) Hydrochloric acid, c(HCl) = 4 mol/l. Add 330 ml of concentrated hydrochloric acid (36%) to about 400 ml water and make up to 1 litre.
(4) Aluminium solution, Al concentration 0.50 mol/l. Dissolve 242 g of aluminium chloride hexahydrate, $AlCl_3 \cdot 6H_2O$, in 2 litre water.
(5) Silver solution. Dissolve 0.85 g of silver nitrate, $AgNO_3$, in 100 ml water.
(6) Stock solution, NO_3 concentration 20 mmol/l. Dissolve 2.0220 g of potassium nitrate, KNO_3, in some water in a 1000-ml volumetric flask and make up to volume.

Standard series

Pipette 0–0.50–1.50–5.0–15–50–150–500 ml of the stock solution (6) into 500-ml volumetric flasks and make up to volume.

This standard series has NO_3 concentrations of 0–0.02–0.06–0.20–0.60–2.0–6.0–20 mmol/l.

Remark:

- Since there are very large differences in nitrate levels in plant material, a wide standard series is necessary. See also the last remark.

Procedure

(a) Preparation of resin

Weigh out 250 g of cation exchange resin (2). Introduce a glass wool tuft down into a percolation tube. Transfer the resin to the percolation tube with the help of water to prevent the formation of air bubbles. Leach dropwise with about 1 litre of 4 M hydrochloric acid (3). Rinse with water until neutral or weakly acid (pH about 5). Then leach the resin dropwise with approximately 1.5 litre of aluminium solution (4) until the pH of the percolate is equal to the pH of the influent (about pH 3). Rinse with water until the leachate is free from chloride; check with silver solution (5). Pour the resin slurry on a Büchner funnel on which a filter paper is put. Suck off the water until the resin looks dry; let the resin dry further on the air.

(b) Measurements

Pipette 5.00 ml of the sample extracts, the blank extracts and the standard series into 50-ml beakers. Add 5.00 ml of buffer solution (1) and a spoonful (0.3–0.6 grams) of resin. Place the beaker on a magnetic stirrer and put a small plastic-coated iron bar in the beaker. Immerse both electrodes in the solution, then start stirring. Read the potential (mV) while stirring; take readings at 0.5-min intervals until three subsequent readings have the same value. Plot a calibration curve on semi-logarithmic paper and read the nitrate concentrations.

Remarks:

- All measurements must be done at constant temperature.
- If an air bubble sticks on the electrode membrane, then simply take the electrode out of the liquid and immerse it again; this will remove the air bubble.
- It is recommended to use A3 size graph paper with a 3-decade logarithmic x-axis and a scale unit of 100 mm per decade.

Calculation

The nitrate content of the dried plant material, expressed in mmol/kg NO_3, is calculated by

$$a \times \frac{V}{w}$$

in which

a = nitrate concentration in the sample extracts, in mmol/l;
V = volume of extractant used for the extraction of w gram sample, in ml;
w = weight of plant material sample, in g.

REFERENCES

I. Novozamsky, V.J.G. Houba, D. van der Eijk and R. van Eck. Notes on determinations of nitrate in plant material. Neth.J.Agric.Sci. *31* (1983) 239–248.

7.2.2 DETERMINATION OF **NITRATE (+ NITRITE)** BY COLORIMETRY (AUTOMATED, BY FLOW ANALYZER)

Field of application

This determination is intended to be carried out on extract 7.0 (water).

The nitrate content of the dried plant material should be higher than 10 mmol/kg.

Principle

In a continuous-flow analysis (CFA) system, the sample is first subjected to dialysis. Nitrate ions from the extract pass the membrane and are taken in an understream of ammonium chloride. The nitrate is then reduced to nitrite by means of copper-coated cadmium. Next, α-naphtylamine and sulphanilamide are added, so that in the acid medium present a red-coloured diazo compound is formed. Its absorbance is measured at a wavelength of 550 nm.

Remarks:

- The dialysis serves to separate nitrate ions from interfering substances like colloids and coloured organic compounds.
- The colour reagent is known as the Griess-Ilosvay reagent.
- Actually, the sum of NO_3 and NO_2 is determined here.

Apparatus

- Continuous-flow analysis system (sampler, pump, dialysis unit, reduction column, NO_3-unit, photometer, recorder). In the authors' laboratory, a Skalar Segmented Flow Analyzer is used, but any CFA system will do.
- Reduction column (a U-shaped glass tubing, about 15 cm long and with internal diameter of 2 mm, provided with ferrules for connection to the CFA tubing). It may be purchased filled with copper-coated cadmium from the CFA system manufacturer, or be home-made (see reagent (5)).

Remark:

- This CFA arrangement allows the determination of nitrite by simply shortcutting the reduction column.

Reagents

(1) Sampler wash fluid. Demineralized water without any wetting agent.
(2) Wetting agent Brij 35, 30% w/v solution, to be purchased from Sigma Chemical Co., product number 430AG-6.
(3) Wash fluid. Add 1 ml of wetting agent (2) to 1 litre of water.
(4) Ammonium chloride solution. Dissolve 10 g of ammonium chloride, NH_4Cl, in some water. Add a little bit (0.5–1 ml) of concentrated aqueous ammonia (25%) so as to bring the pH to 6–6.5, and make up to 1 litre. Add 1 ml of wetting agent (2).
(5) Cd-Cu reductor. Swirl approximately 5 g of cadmium powder (particle size 0.3–0.8 mm) during 1 min with about 30 ml of 1 M hydrochloric acid. Wash with water until acid-free. Then add about 50 ml of a copper(II) sulphate solution (20 g/l) and swirl during 3 min. Wash at least 10 times with water to remove any flocculated copper.
Fill the U-shaped column with NH_4Cl, taking care not to introduce air bubbles. Introduce the activated Cd powder with the aid of a funnel on both sides of the column. Apply vibration now and then to pack the column. Fill the column up to 5 mm from the top and seal the ends with a small plug of glass wool. The column is now ready for use and can be placed in the CFA system.
Store the activated Cd/Cu powder in the reduction column in a dark place.
(6) Colour reagent. Add 100 ml of concentrated phosphoric acid (85%) to about 400 ml of water in a 1000-ml volumetric flask. Add 0.5 g of α-naphtylamine, $C_{10}H_9N$, and swirl until dissolved. Then dissolve 10 g of sulphanilamide, $C_6H_8N_2O_2S$, in this mixture. Add 0.5 ml of wetting agent (2) and make up to the mark.
(7) Standard solution, NO_3 concentration 3000 mg/l. Dissolve 4.892 g of potassium nitrate, KNO_3, in water in a 1000-ml volumetric flask and make up to volume.

Standard series

Pipette into 100-ml volumetric flasks 0–3.33–6.67–10.00 ml of the standard solution (6) and make up to volume. This standard series has NO_3 concentrations of 0–100–200–300 mg/l.

Procedure

Make up a measuring series by placing standards and samples in the automatic sampler in the following sequence:

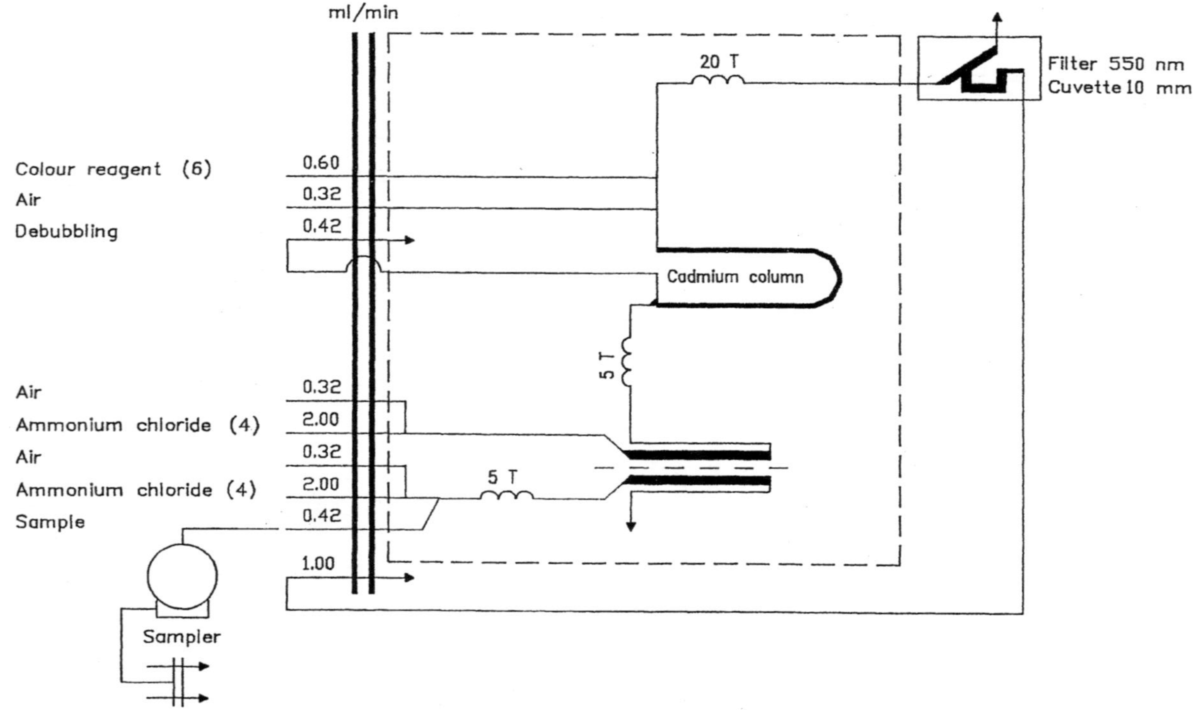

Flow diagram NO_3-determination (7.2.2).

1–3 fourth term of the standard series (highest standard)
4 first term of the standard series (zero standard)
5 second term of the standard series
6 third term of the standard series
7 fourth term of the standard series (highest standard)
8 first term of the standard series (zero standard)
9 blank extraction
10–48 rest of the extraction series (i.e., 34 samples, two standard samples "low", two standard samples "high" and one more blank extraction)
49–50 fourth term of the standard series (highest standard).

Put the continuous-flow system into operation according to the scheme given herewith. Rinse it during 5 minutes with wash fluid (3) and then for 30 minutes with reagents. Then put the sampler into operation and measure the absorbance at a wavelength of 550 nm. Plot a calibration curve and read the nitrate concentrations.

After finishing the measurements (at the end of the day), shortcut the reduction column and flush the system during 30 min with wash fluid (3).

Remarks:

- The highest standards at positions 1 and 2 serve to calibrate the recorder at a desired peak height. The standard at position 3, then, must give an equally high peak as the last two standards (49 and 50).
- The zero standard (position 8) may give too high a result because of influence of the highest standard measured just before. Therefore, this observation is rejected.

Calculation

The nitrate content of the dried plant material, expressed in mmol/kg NO_3, is calculated by

$$0.0161 \times (a - b) \times \frac{V}{w}$$

in which
a = concentration of nitrate in the sample extract, in mg/l;
b = concentration of nitrate in the blank extract, in mg/l;
V = volume of water used for extraction, in ml;
w = weight of plant material sample, in g.

7.3 DETERMINATION OF **NITRITE** BY COLORIMETRY (AUTOMATED, BY FLOW ANALYZER)

Field of application

This determination is intended to be carried out on extract 7.0 (water).

The nitrite content of the dried plant material should be higher than 2 mmol/kg.

Principle

In a continuous-flow analysis (CFA) system, the sample is first subjected to dialysis. Nitrite ions from the extract pass the membrane and are taken in an understream of ammonium chloride. By addition of α-naphtylamine and sulphanilamide in acid medium, the nitrite forms a red-coloured diazo compound, of which the absorbance is measured at a wavelength of 550 nm.

Remarks:
- The dialysis serves to separate ions from interfering substances like colloids and coloured organic compounds.
- The colour reagent is known as the Griess-Ilosvay reagent.

Apparatus

- Continuous-flow analysis system (sampler, pump, dialysis unit, NO_2-unit, recorder). In the authors' laboratory, a Skalar Segmented Flow Analyzer is used, but any CFA system will do.

Remark:
- In the authors' laboratory the CFA arrangement of determination 7.2.2 (nitrate + nitrite) is used by simply shortcutting the Cu-Cd reduction column.

Reagents

(1) Sampler wash fluid. Demineralised water without any wetting agent.
(2) Wetting agent Brij 35, 30% w/v solution, to be purchased from Sigma Chemical Co., product number 430AG-6.
(3) Wash fluid. Add 1 ml of wetting agent (2) to 1 litre of water.
(4) Ammonium chloride solution. Dissolve 10 g of ammonium chloride, NH_4Cl, in 1 litre of water and add 1 ml of wetting agent (2).
(5) Colour reagent. Add 100 ml of concentrated phosphoric acid (85%) to about 600 ml of water in a 1000-ml volumetric flask. Add 0.5 g of α-naphtylamine, $C_{10}H_9N$, and swirl until dissolved. Then dissolve 10 g of sulphanilamide, $C_6H_8N_2O_2S$, in this mixture. Add 0.5 ml of wetting agent (2) and make up to the mark.
(6) Standard solution, NO_2 concentration 300 mg/l. Dissolve 0.4499 g of sodium nitrite, $NaNO_2$, in water in a 1000-ml volumetric flask and make up to volume.

Standard series

Pipette into 100-ml volumetric flasks 0–3.33–6.67–10.00 ml of the standard solution (5) and make up to volume. This standard series has NO_2 concentrations of 0–10–20–30 mg/l.

Procedure

Make up a measuring series by placing standards and samples in the automatic sampler in the following sequence:

1–3 fourth term of the standard series (highest standard)
4 first term of the standard series (zero standard)
5 second term of the standard series
6 third term of the standard series
7 fourth term of the standard series (highest standard)
8 first term of the standard series (zero standard)
9 blank extraction
10–48 rest of the extraction series (i.e., 34 samples, two standard samples "low", two standard samples "high" and one more blank extraction)
49–50 fourth term of the standard series (highest standard).

Put the continuous-flow system into operation according to the scheme given herewith. Rinse it during 5 minutes with wash fluid (3) and then for 30 minutes with reagents; then put the sampler into operation. Measure the absorbance at a wavelength of 550 nm. Plot a calibration curve and read the nitrite concentrations.

After finishing the measurements (at the end of the day), flush the system during 30 min with wash fluid (3).

Remarks:

- The highest standards at positions 1 and 2 serve to calibrate the recorder at a desired peak height. The standard at position 3, then, must give an equally high peak as the last two standards (49 and 50).
- The zero standard (position 8) may give too high a result because of influence of the highest standard measured just before. Therefore, this observation is rejected.

Calculation

The nitrite content of the dried plant material, expressed in mmol/kg NO_2, is calculated by

$$0.0217 \times (a - b) \times \frac{V}{w}$$

in which

a = concentration of nitrite in the sample extract, in mg/l;
b = concentration of nitrite in the blank extract, in mg/l;
V = volume of water used for extraction, in ml;
w = weight of plant material sample, in g.

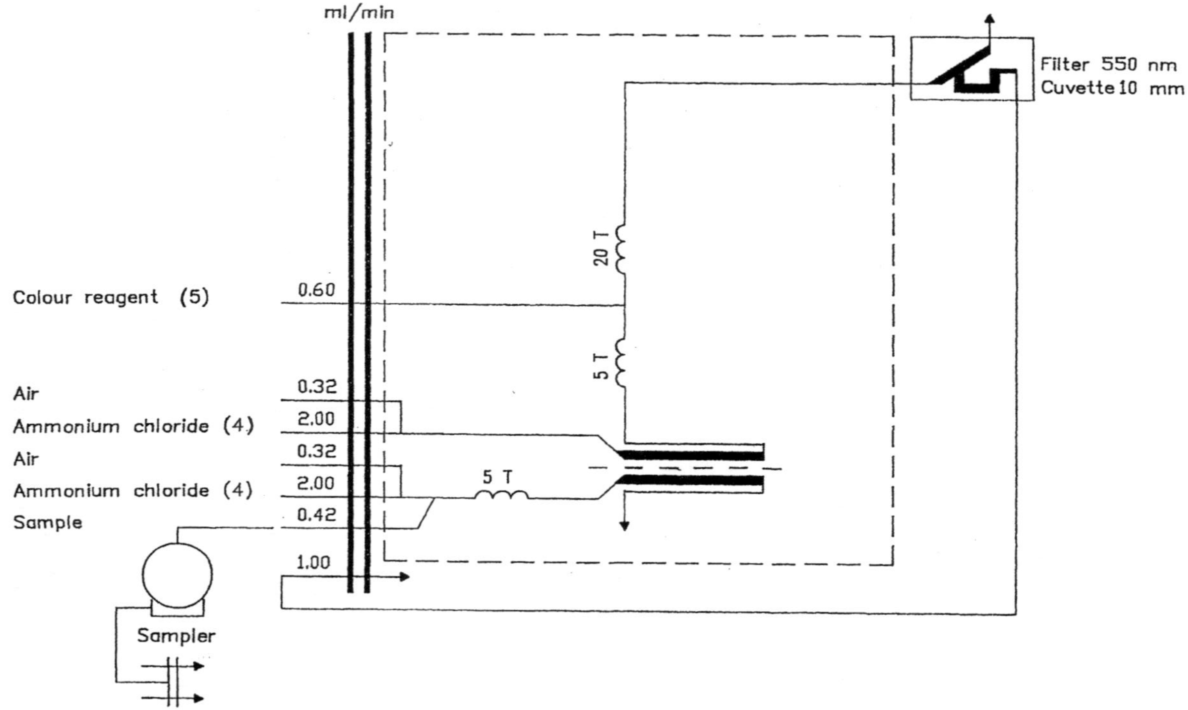

Flow diagram NO_2-determination (7.3).

7.4 DETERMINATION OF **SULPHATE** BY ICP-ES

Field of application

This determination is to be carried out on extract 7.0 (water).

The sulphate content of the dried plant material should be higher than 25 mmol/kg.

Principle

Sulphate ions are precipitated by barium ions. After clean-up, the barium sulphate precipitate is dissolved by EDTA. The solution is then nebulized into an argon plasma, where all components are vaporized. Sulphate ions are broken down to sulphur atoms, which are excited and then emit radiation of which the intensity is measured at a wavelength of 182.04 nm.

Remark:

- The short wavelength requires a monochromator that can be evacuated or flushed with nitrogen gas to remove interfering oxygen.

Apparatus

- Inductively-coupled plasma emission spectrometer, with a monochromator that can either be evacuated or flushed with an inert gas.

Reagents

(1) Barium chloride solution, c($BaCl_2$) = 1 mol/l. Dissolve 244 g of barium chloride dihydrate, $BaCl_2 \cdot 2H_2O$, in about 600 ml of water. Add 82 ml of concentrated hydrochloric acid (36%) and make up to 1 litre.

(2) EDTA solution, c(H_4EDTA) = 0.02 mol/l. Dissolve 5.84 g of ethylene diamine tetra acetic acid, H_4EDTA, in 30 ml of concentrated aqueous ammonia (25%). Make up to 1 litre with water.

(3) Stock solution, S concentration 1250 mg/l. Dissolve 6.793 g of potassium sulphate, K_2SO_4, in some water in a 1000-ml volumetric flask. Make up to volume with EDTA solution (2).

Standard series

Pipette 0–4.00–8.00–12.00 ml of the stock solution (3) into 100-ml volumetric flasks. Make up to the mark with EDTA solution (2). This standard series has S concentrations of 0–50–100–150 mg/l.

Procedure

(a) Separation of sulphate

Weigh a series of 15-ml centrifuge tubes (empty weight e gram per tube). Add to each tube 1.0 ml of barium chloride solution (1). Then pipette 10.0 ml of the extracts and the standard series into these tubes, and swirl to mix. Bring the tubes two by two with water at the same weight, and centrifuge for 10 min at 2000 **g**. Carefully decant the supernatant, so that the precipitate and only about 0.5 ml of liquid will remain in the tube. Next, add 5.0 ml of water and repeat the centrifuging. Decant again the supernatant and weigh the tube (c gram) in order to establish the volume of liquid left behind. Finally, add 5.0 ml of EDTA solution (2) and swirl to bring the precipitate into suspension. Allow the precipitate to dissolve completely, preferably overnight but at least during 2 hours; swirl now and then.

(b) Calibration of the spectrometer

Flush the monochromator with nitrogen gas for 2 hours. Then ignite the plasma and continue the flushing until the measurements have been finished. Allow the plasma to stabilize for at least one hour before the measurements will be made.

Set the wavelength of the spectrometer at 182.04 nm; fix the other operating parameters (power, gas flows, observation height, integration time, read delay etc.) at their – previously elaborated – appropriate values for sulphur determination.

Nebulize, in this order, the highest standard (150 mg/l) and the zero standard, and enter their nominal concentration values. The spectrometer is now calibrated in concentration units.

Next, measure the concentrations of the highest standard and the zero standard as if it were samples, to check the calibration procedure. When the values thus found deviate more than 10% from the nominal values, the calibration procedure should be repeated. In the other case, measure the concentration of the rest of the standard series to check the required linearity.

(c) Measurement

Nebulize the sample extracts and the blank extracts and measure the sulphur concentrations immediately after calibration. With intervals of 9 samples, measure the highest standard and the zero standard again. Correct the results for the samples arithmetically for any zero drift or change in sensitivity.

Remarks:

- In the course of the day, the sensitivity changes continuously. With the authors' apparatus this change is seldom larger than 10% over a 9-sample interval. In this case, the arithmetical correction procedure is better than recalibration.
- No interference was observed when nebulizing 1000 mg/l-solutions of Al, As, Ba, Ca, Cd, Cr, Cu, Fe, K, Mg, Mn, Na, Ni, P, Sr and Ti, and 100 mg/l-solutions of B and Mo.
- In the authors' laboratory, the limit of determination (for 5% C.V., based on quadruplate instrument readings) proved to be 2.7 mg/l, corresponding with a sulphate content of 25 mmol/kg dry plant material. This is sufficiently low for almost any plant material.

Calculation

The sulphate content of the dried plant material, expressed in mmol/kg SO_4, is calculated by

$$\frac{(c-e) + 5}{5} \times 0.0156 \times (a-b) \times \frac{V}{w}$$

in which

c = weight of centrifuge tube with remaining liquid, in g;
e = weight of empty centrifuge tube, in g;
a = concentration of sulphate in the sample extract, in mg/l;
b = concentration of sulphate in the blank extract, in mg/l;
V = total volume of water taken for extraction, in ml;
w = weight of plant material sample, in g.

REFERENCES

I. Novozamsky, R. van Eck, J.J. van der Lee, V.J.G. Houba and E. Temminghoff. Determination of total sulphur and extractable sulphate in plant materials by inductively-coupled plasma atomic emission spectrometry. Commun. Soil Sci. Plant Anal. *17* (1986) 1147–1157.

8 Extraction with **water** in the presence of Ag and Cu and determination of NO_3

8.0 Extraction with **water** in the presence of Ag and Cu
8.1 Determination of **nitrate** by colorimetry after distillation

8.0 EXTRACTION WITH **WATER** IN THE PRESENCE OF Ag AND Cu

Field of application

This extraction is meant exclusively for the determination of NO_3^- by the o-xylenol method.

Principle

The nitrate that is "free" in the plant material is extracted as such by water.

Remarks:

- Silver sulphate is added to the extractant to bind chloride, oxalate and tannin, which would otherwise interfere.
- Copper sulphate is added to prevent growth of micro-organisms.

Apparatus

- Shaker, linear or end-over-end.

Reagents

(1) Extracting solution. Dissolve 7.0 g of silver sulphate, Ag_2SO_4, and 40 g of copper(II) sulphate pentahydrate, $CuSO_4 \cdot 5H_2O$, in water and make up to 1 litre with water. Stir at least for one hour to dissolve the silver sulphate as far as possible. Filter.

Procedure

Weigh, to the nearest 0.001 g, an amount of w gram (w varying between 0.5 and 3 g, depending on the expected amount of NO_3^-) of the dried plant material sample in a 100-ml erlenmeyer flask. Add 50.0 ml of extracting solution (1) and shake mechanically during 30 min. Filter over medium fine paper. Prepare also a blank extract.

Remark:

- Since no chloride interference has to be feared, a maximum amount of 3 g of plant material is allowed even with Cl^- contents as high as 3%.

REFERENCES

C.Sorensen. The xylenol method and determination of nitrate in beets. Physiol. Plantarum *9* (1956) 304–320.

8.1 DETERMINATION OF **NITRATE** BY COLORIMETRY AFTER DISTILLATION

Field of application

This determination is to be carried out on extract 8.0 (water + Ag + Cu).

Principle

In strong sulphuric acid medium, nitrate reacts with o-xylenol to form nitro-xylenol. This compound is distilled over with ethanol into sodium hydroxide solution, where it dissolves yielding a yellow colour. Its absorbance is measured at a wavelength of 430 nm.

Remark:

- The distillation is carried out in alcoholic atmosphere to prevent condensation of nitro-xylenol in the still.

Apparatus

- Steam generator, electrically heated (see Fig. 1).
- Distillation unit ("Markham still"), micro or semi-micro design (see Fig. 2).

Remarks:

- This steam generator is protected against overloading; the production of steam is adjusted automatically according to the consumption of steam. In the authors' laboratory, a Büchi DG-1500 steam generator is used (manufactured by Büchi, CH-9230 Flawil, Switzerland).
- This still has the evaporator vessel enclosed in a combined steam jacket and trap; an arrangement which is compact and accelerates distillation. In the authors' laboratory, a Quickfit 46 MC distillation unit is used (manufactured by Corning Ltd., Laboratory Division, Stone, Staffordshire ST15 0BG, England).

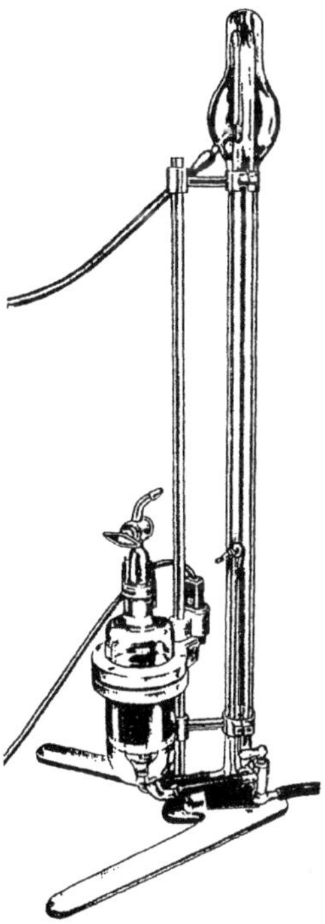

Fig. 1. Steam generator.

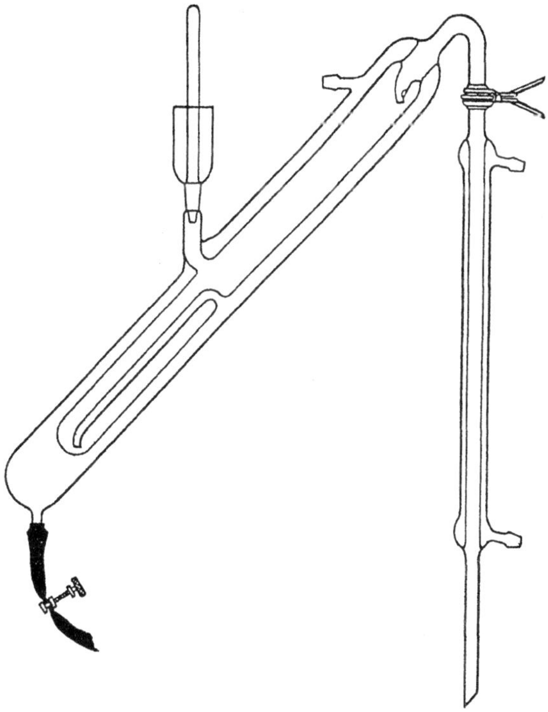

Fig. 2. Distillation unit.

Reagents

(1) Sulphuric acid, $c(H_2SO_4)$ = 14.4 mol/l. While swirling and cooling, add carefully, in small portions, 720 ml of concentrated sulphuric acid (96%) to 280 ml of water. (The end volume will be 900 ml.)
(2) Sodium hydroxide solution, c(NaOH) = 0.4 mol/l. Dissolve 16 g of sodium hydroxide, NaOH, in 1 litre of water.
(3) o-Xylenol solution. Dissolve 2.5 g of ortho-xylenol (3,4-dimethylphenol), $C_8H_{10}O$, in 50 ml of 0.4 M sodium hydroxide solution (2) and dilute to 100 ml with water.
(4) Ethanol, 96% (v/v).
(5) Extracting solution. Dissolve 7.0 g of silver sulphate, Ag_2SO_4, and 40 g of copper(II) sulphate pentahydrate, $CuSO_4 \cdot 5H_2O$, in water and make up to 1 litre with water. (See section 8.0)
(6) Stock solution, NO_3 concentration 200 mg/l. Dissolve 0.3261 g of potassium nitrate, KNO_3, in water in a 1000-ml volumetric flask and make up to volume with water.

Standard series

Pipette 0–1.00–2.00–3.00–5.00 ml of the stock solution (6) into 100-ml erlenmeyer flasks with ground necks. Dilute with water to exactly 5.00 ml. This standard series has NO_3 concentrations of 0–40–80–120–200 mg/l.

Procedure

a) Nitration

Pipette D ml (D varying between 1.00 and 5.00 ml, depending on the expected amount of NO_3^- of the extract) into ground-stoppered erlenmeyer flasks of 100 ml. Make up to 5.00 ml with the extracting solution (5). Add to the sample extracts, the blank extract and the standard series first 2.0 ml of o-xylenol solution (3) and then 20.0 ml of sulphuric acid (1). Stopper immediately. Swirl and leave to stand for 30 min; then add 50.0 ml water and swirl until homogeneous.

b) Distillation

Pipette 25.0 ml of the nitrated solution from a) into the distillation unit; add 5 ml of ethanol (4) through the funnel. Distil during 1.5 min; collect the distillate in a 25-ml volumetric flask containing 2.5 ml of 0.4 M sodium hydroxide solution (2). While distilling, add again 5 ml of ethanol (4), taking care that no vapours escape by closing the cock before the last drop of ethanol runs in, and proceed with the distillation until the volumetric flask is nearly filled. Stop distilling, make up to the mark with ethanol (4) and mix.

c) Measurement

Measure the absorbance of all distillates in a 1-cm cuvette at a wavelength of 430 nm. Plot a calibration curve and read the nitrate concentrations.

Remarks:

- The calibration curve should be linear.
- The volume of sample extract taken for determination (= D ml) should not contain more than 1 mg of NO_3^-.
- Protect the solutions during and after nitration against direct sunlight, because otherwise the solutions will become very dark and erroneous results will be obtained.
- The time of nitration is not critical between 10 and 60 minutes. After one hour, however, the strong sulphuric acid will start to break down the o-xylenol. Addition of water dilutes the acid sufficiently to prevent decomposition. Thereafter, the solution may be left to stand for 6 hours.
- No effect was observed from variations in ambient temperature between 13 and 35 °C.
- Steam distillation is preferred here, since with a "normal" distillation the sulphuric acid concentration will gradually increase. Liquid-liquid extraction of nitro-xylenol has been tried, but then deterioration products of sugars (originally present in the plant sample) cause interference.
- Do not use rubber stoppers, since sulphuric acid attacks the stoppers.
- When using the right xylenol isomer (i.e., 3,4-dimethylphenol), the method is virtually specific.
- Nitrite causes interference, but can be removed by addition of sulfamic acid (NH_2SO_3H).

- No interferences have been found when spiking 5 ml of extracting solution with 100 mg Cu(II), 100 mg Mn(II), 100 mg NH_4, 100 mg Ni(II), 75 mg Al, 50 mg Sn(II) or 50 mg Sn(IV).
- In the presence of Fe(III) an iron-nitroxylenol compound may precipitate and too low NO_3— values may be found.
- If no nitrate is present, vanadate shows a positive interference. More than 10 mg of vanadate per 5 ml of extracting solution causes excessive foaming, making distillation impossible.
- Organic substances may cause overdistillation of foam, but addition of antifoam may help to reduce the foaming.

Calculation

The nitrate content of the dried plant material, expressed in mmol/kg NO_3, is calculated by

$$0.0806 \times (a - b) \times \frac{1}{D} \times \frac{V}{w}$$

in which
a = concentration of nitrate in the sample extract, in mg/l;
b = concentration of nitrate in the blank extract, in mg/l;
D = volume of extract taken for determination, in ml;
V = total volume of extract, in ml;
w = weight of plant material sample, in g.

REFERENCES

I. Novozamsky, V.J.G. Houba, D. van der Eijk and R. van Eck. Notes on determinations of nitrate in plant material. Neth.J.agric.Sci. *31* (1983) 239–248.

9 Extraction with HF-HCl and determination of B, Si

9.0 EXTRACTION WITH HF-HCl

Field of application

This extraction is only appropriate for the determination of B and Si.

Principle

Borates and silicates are extracted from the plant material as tetrafluoroborates and hexafluorosilicates, respectively.

Remarks:
- If a sample contains both Si and K in large concentrations, the slightly soluble K_2SiF_6 may be formed. A larger extraction ratio should then be used.
- The extracts are coloured brown, so that colorimetric determinations are not possible.

Apparatus

- End-over-end shaker.
- Polycarbonate test tubes and caps.
- Polythene funnels.
- Polythene volumetric flasks.
- Polythene graduated cylinders.
- Pipettors with polythene tips.

Reagents

(1) Hydrofluoric acid, 40% (w/w), c(HF) = 22.6 mol/l (d = 1.13 g/ml).
(2) Hydrochloric acid, 36% (w/w), c(HCl) = 12.0 mol/l (d = 1.19 g/ml).
(3) Aqua bidest.

(4) Extraction mixture. Mix 100 ml of hydrofluoric acid (1) with 40 ml of hydrochloric acid (2) and add 360 ml aqua bidest (3). This extraction solution is 4.5 M in HF and 1 M in HCl.

Procedure

Weigh, to the nearest 0.001 g, approximately 0.3 g of the dried plant material sample into a polycarbonate test tube. Add 10 ml of extraction mixture (4) and moisten the plant material by shaking manually. Prepare also a blank extraction.

Shake overnight at ambient temperature (20 ± 2°C). Filter the suspension over medium fine paper.

Remarks:

- Since normal demineralized water may contain varying concentrations of Si, the use of doubly distilled water is strongly advised. Instead of this aqua bidest, the common demineralized water may be distilled once.
- Hydrofluoric acid is a treacherous skin poison. Wear rubber gloves, safety goggles and protective cloths. If any HF comes into contact with the skin, wash immediately and thoroughly with water and thereafter dab with calcium gluconate gel.

REFERENCES

I. Novozamsky, R. van Eck and V.J.G. Houba. A rapid determination of silicon in plant material. Commun. Soil Sci. Plant Anal. *15* (1984) 205–211.

J.J. van der Lee, I. Walinga, P.K. Manyeki, V.J.G. Houba and I. Novozamsky. Determination of boron in fresh and in dried plant material by plasma emission spectrometry after extraction with HF-HCl. Commun. Soil Sci. Plant Anal. *18* (1987) 789–802.

9.1 DETERMINATION OF **BORON** BY ICP-ES

Field of application

This determination is intended to be carried out on extract 9.0 (HF-HCl).

The boron content of the dried plant material should be higher than 2.5 mg/kg.

Principle

Solutions with boron compounds are nebulized into an argon plasma, where all components are vaporized. Boron compounds are dissociated and excitated, and then emit radiation of which the intensity is measured at a wavelength of 249.77 nm.

Remark:

- High concentrations of Fe cause spectral interference.

Apparatus

- End-over-end shaker.
- Polycarbonate test tubes.
- Polythene bottles.
- Polythene volumetric flasks.
- Inductively coupled plasma emission spectrometer, fitted with HF-resistant nebulizer and inner torch tube.

Reagents

(1) Extracting solution. Mix 150 ml of concentrated hydrofluoric acid (40%) with 60 ml of concentrated hydrochloric acid (36%); add 540 ml water. (This solution is 4.5 M in HF and 1.0 M in HCl; it is identical to the extraction solution of section 9.0).

(2) Stock solution, B concentration 100 mg/l. Dissolve 0.5719 g of boric acid, H_3BO_3, in some water in a 1000-ml volumetric flask and make up to volume. Transfer immediately to a polythene bottle.

(3) Standard solution, B concentration 20 mg/l. Pipette 20.0 ml of the stock solution (2) to a 100-ml polythene volumetric flask. Add 20 ml of concentrated hydrofluoric acid (40%) and 8 ml of concentrated hydrochloric acid (36%) and dilute to volume.

Remark:
- The pipetting should be done by means of piston-type pipettes with polythene tips.

Standard series

Pipette 0–1.00–2.00–3.00–4.00–5.00 ml of the standard solution (3) into 100-ml polythene volumetric flasks. Make up to the mark with extracting solution (1). This standard series has B concentrations of 0–0.2–0.4–0.6–0.8–1.0 mg/l.

Procedure

(a) Calibration of the spectrometer
Set the wavelength of the spectrometer at 249.77 nm; set the other operating parameters (power, gas flows, observation height, integration time, read delay etc.) at their – previously elaborated – appropriate values for boron determination.

Nebulize, in this order, the highest standard (1.0 mg/l) and the zero standard and enter their nominal concentration values. The spectrometer is now calibrated in concentration units.

Next, measure the concentration of the highest standard and the zero standard as if it were samples, to check the calibration procedure. When the values thus found deviate more than 10% from the nominal values, the calibration procedure should be repeated. In the other case, measure the concentration of the rest of the standard series to check the required linearity.

(b) Measurement of the samples

Nebulize the extracts and the blank extracts and measure the boron concentrations at a wavelength of 249.77 nm and otherwise the same measuring conditions at which the spectrometer was calibrated. After intervals of 5–6 samples, measure the highest standard and the zero standard again. Correct the results for the samples arithmetically for any zero drift or change in sensitivity.

Remarks:

- In the course of the day, the sensitivity changes with time like a wave: alternating small decreases and increases without significant drift. Therefore, the arithmetical correction procedure is better than recalibration, as long as the sensitivity change is not too large (say 10%).
- This determination can also be applied to fresh material (see literature reference).

Calculation

The boron content of the dried plant material, expressed in mg/kg B, is calculated by

$$(a - b) \times \frac{V}{w}$$

in which

a = concentration of boron in the sample extract, in mg/l;
b = concentration of boron in the blank extract, in mg/l;
V = volume of acid mixture used for extraction, in ml;
w = weight of plant material sample, in g.

REFERENCES

J.J. van der Lee, I. Walinga, P.K. Manyeki, V.J.G. Houba and I. Novozamsky. Determination of boron in fresh and in dried plant material by plasma emission spectrometry after extraction with HF-HCl. Commun. Soil Sci. Plant Anal. *18* (1987) 789–802.

9.2.1 DETERMINATION OF **SILICON** BY ICP-ES

Field of application

This determination is intended to be carried out on extract 9.0 (HF-HCl).

The silicon content of the dried plant material should be higher than 500 mg/kg.

Principle

Solutions with silicon compounds are nebulized into an argon plasma, where all components are vaporized. Silicon compounds are dissociated and excited, and then emit radiation of which the intensity is measured at a wavelength of 251.61 nm.

Apparatus

- Polycarbonate test tubes + stoppers.
- Polythene bottles.
- Polythene volumetric flasks.
- Inductively-coupled plasma emission spectrometer, fitted with HF-resistant nebulizer and inner torch tube.

Reagents

For this determination, all solutions must be prepared with doubly distilled water *(bidest)*, since demineralized water may contain varying amounts of Si.

(1) Acid mixture. Mix 150 ml of concentrated hydrofluoric acid (40%) with 60 ml of concentrated hydrochloric acid (36%) and add 540 ml bidest. (This solution is 4.5 M in HF and 1.0 M in HCl; it is identical to the extraction solution of section 9.0)

Standard series

Dissolve 0–26.7–53.5–106.9–160.4–213.9–267.4 mg of silica, SiO_2, finely ground, dried at 130 °C, in the acid mixture (1) in 250-ml polythene volumetric flasks. If dissolving proceeds slowly, close the flasks and warm them in a stove at a maximum temperature of 65°C. After cooling down to room temperature, make up to the mark with acid mixture (1). This standard series has Si concentrations of 0–50–100–200–300–400–500 mg/l.

Remarks:

- The usual way of making a standard series by diluting a stock solution is not recommended, because the necessary concentration of the stock solution would be too high for complete dissolution. A less concentrated stock solution would require large-volume HF-resistant pipettes, which may not be available.
  Alternatively, incorporating the required 1:20 dilution (see Procedure) in preparing the standard series would imply the use of small pipettes; in that case, however, the standard series would not be diluted in exactly the same way as the samples.
- The dilution for the standards and samples must be done with bidest of one batch, i.e., having the same quality.

Procedure

(a) Dilution
Using piston-type pipettes or a diluting apparatus with polythene tips, dilute 0.50 ml of the standard series, the sample extracts and the blank extracts with 9.50 ml of bidest in polycarbonate test tubes. This diluted standard series has Si concentrations of 0–2.5–5–10–15–20–25 mg/l.

(b) Calibration of the spectrometer
Set the wavelength of the spectrometer at 251.61 nm; fix the other operating parameters (power, gas flows, observation height, integration time, read delay etc.) at their – previously elaborated – appropriate values for silicon determination.

Nebulize, in this order, the diluted highest standard (25 mg/l) and the diluted zero standard, and enter their nominal diluted concentration values. The spectrometer is now calibrated in concentration units.

Next, measure the concentrations of the diluted highest standard and the diluted zero standard as if it were samples, to check the calibration procedure. When the values thus found deviate more than 10% from the nominal values, the calibration procedure should be repeated. In the other case, measure the concentration of the rest of the standard series to check the required linearity.

(c) Measurement
Nebulize the diluted extracts and measure the silicon concentrations immediately after calibration. With intervals of 5–6 samples, measure the diluted highest standard and the diluted zero standard again. Correct the results for the samples arithmetically for any zero drift or change in sensitivity.

Remarks:

- In the course of the day, the sensitivity changes with time like a wave: alternating small decreases and increases without significant drift. Therefore, the arithmetical correction procedure is better than recalibration, as long as the sensitivity change is not too large (say 10%).
- New polycarbonate test tubes, pipette tips etc. must be cleaned by rinsing with the acid mixture (1), because new material may release silicon.
- Samples with Si concentrations beyond the standard series should be diluted first with acid mixture (1) and then follow the procedure with the usual 1:20 dilution.

Calculation

The silicon content of the dried plant material, expressed in mg/kg Si, is calculated by

$$20 \times (a - b) \times \frac{V}{w}$$

in which
a = concentration of silicon in the diluted sample extract, in mg/l;

b = concentration of silicon in the diluted blank extract, in mg/l;
V = volume of acid mixture used for extraction, in ml;
w = weight of plant material sample, in g.

REFERENCES

I. Novozamsky, R. van Eck and V.J.G. Houba. A rapid determination of silicon in plant material. Commun. Soil Sci. Plant Anal. *15* (1984) 205–211.

9.2.2 DETERMINATION OF **SILICON** BY FLAME AAS

Field of application

This determination is intended to be carried out on extract 9.0 (HF-HCl).

The silicon content of the dried plant material should be higher than 500 mg/kg.

The ICP method (procedure 9.2.1) is preferred, because it allows a twentyfold dilution with water, which lowers the vapour pressure of the hydrochloric acid appreciably.

Principle

The sample is nebulized into a reducing nitrous oxide-acetylene flame, where it is vaporized. Silicon compounds are atomized, and the silicon atoms thus formed absorb radiation from a hollow-cathode lamp. The absorbance is measured at a wavelength of 251.6 nm.

Apparatus

- Polycarbonate test tubes + stoppers.
- Polythene bottles.
- Polythene volumetric flasks.
- Atomic absorption spectrophotometer, fitted with a nitrous oxide-acetylene burner.

Reagents

For this determination, all solutions must be prepared with doubly distilled water *(bidest)*, since demineralized water may contain varying amounts of Si.

(1) Acid mixture. Mix 150 ml of concentrated hydrofluoric acid (40%) with 60 ml of concentrated hydrochloric acid (36%) and add 540 ml bidest. (This solution is 4.5 M in HF and 1.0 M in HCl; it is identical to the extraction solution of section 9.0)

Standard series

Dissolve 0–26.7–53.5–106.9–160.4–213.9–267.4 mg of silica, SiO_2, finely ground, dried at 130 °C, in the acid mixture (1) in 250-ml polythene volumetric flasks. If dissolving proceeds slowly, close the flasks and warm them in a stove at a maximum temperature of 65°C. After cooling down to room temperature, make up to the mark with acid mixture (1). This standard series has Si concentrations of 0–50–100–200–300–400–500 mg/l.

Remarks:
- The usual way of making a standard series by diluting a stock solution is not recommended, because the necessary concentration of the stock solution would be too high for complete dissolution. A less concentrated stock solution would require large-volume HF-resistant pipettes, which may not be available.
- Hydrofluoric acid causes a severe depression of the silicon atomic absorption signal; therefore a standard series with relatively high Si concentrations is used.

Procedure

Nebulize the sample extracts, the blank extracts and the standard series into a red-coloured (reducing) nitrous oxide-acetylene flame and measure the absorbance at a wavelength of 251.6 nm. Plot a calibration curve and read the silicon concentrations.

Remarks:
- Working with a nitrous oxide-acetylene flame is more dangerous than with an air-acetylene flame. Therefore, extra care should be taken to apply the right procedure for nitrous oxide-acetylene flames.
- The position of the burner is rather critical for the nitrous oxide-acetylene flame. Therefore, both the burner height and the horizontal burner alignment should be optimized while nebulizing a Si standard.
- New polycarbonate test tubes, pipette tips etc. must be cleaned by rinsing with the acid mixture (1), because new material may release silicon.
- No interference was observed when spiking the measuring solutions to give concentrations of 800 mg/l Al, and 1000 mg/l of Ca, F, Na and P in the presence of 100 mg/l Si.

Calculation

The silicon content of the dried plant material, expressed in mg/kg Si, is calculated by

$$(a - b) \times \frac{V}{w}$$

in which
a = concentration of silicon in the sample extract, in mg/l;
b = concentration of silicon in the blank extract, in mg/l;
V = volume of acid mixture used for extraction, in ml;
w = weight of plant material sample, in g.

REFERENCES

I. Novozamsky, R. van Eck and V.J.G. Houba. A rapid determination of silicon in plant material. Commun. Soil Sci. Plant Anal. *15* (1984) 205–211.

Appendices

APPENDIX A-1. ATOMIC WEIGHTS

The atomic weights used throughout these appendices are the standard atomic weights 1987 as given by the IUPAC, scaled to ^{12}C = 12. The following table gives these atomic weight values only for the elements which appear in the syllabi and in the present appendices. The values are given with the same number of decimals as in the original IUPAC table, i.e. with the same degree of confidence. The last column gives the uncertainty of the last decimal digit. The values apply to elements as they exist naturally on earth, and are thus applicable to any normal material.

Name	Symbol	Atomic number	Atomic weight	
silver	Ag	47	107.8682	± 2
aluminium	Al	13	26.981539	± 5
arsenic	As	33	74.92159	± 2
boron	B	5	10.811	± 5
barium	Ba	56	137.327	± 7
bismuth	Bi	83	208.98037	± 3
bromine	Br	35	79.904	± 1
carbon	C	12	12.011	± 1
calcium	Ca	20	40.078	± 4
cadmium	Cd	48	112.411	± 8
cerium	Ce	58	140.115	± 4
chlorine	Cl	17	35.4527	± 9
cobalt	Co	27	58.93320	± 1
chromium	Cr	24	51.9961	± 6
cesium	Cs	55	132.90543	± 5
copper	Cu	29	63.546	± 3
fluorine	F	9	18.9984032	± 9
iron	Fe	26	55.847	± 3
hydrogen	H	1	1.00794	± 7
mercury	Hg	80	200.59	± 3
iodine	I	53	126.90447	± 3
potassium	K	19	39.0983	± 1
lanthanum	La	57	138.9055	± 2
lithium	Li	3	6.941	± 2
magnesium	Mg	12	24.3050	± 6
manganese	Mn	25	54.93805	± 1
molybdenum	Mo	42	95.94	± 1
nitrogen	N	7	14.00674	± 7
sodium	Na	11	22.989768	± 6
nickel	Ni	28	58.69	± 1
oxygen	O	8	15.9994	± 3
phosphorus	P	15	30.973762	± 4
lead	Pb	82	207.2	± 1
sulphur	S	16	32.066	± 6
antimony	Sb	51	121.75	± 3
selenium	Se	34	78.96	± 3
silicon	Si	14	28.0855	± 3
tin	Sn	50	118.710	± 7
strontium	Sr	38	87.62	± 1
titanium	Ti	22	47.88	± 3
zinc	Zn	30	65.39	± 2

REFERENCES

IUPAC commission on Atomic Weights and Isotopic Abundances: Atomic Weights of the Elements 1987. Pure Appl. Chem. *60* (1988) 841–854.

APPENDIX A-2. OLD AND NEW (SI) UNITS

This table gives a survey of the quantities and units most frequently used in the soil and plant analysis laboratory. Being far from complete, it is only meant as an aid to memory for the most important units and their conversion. More information can be found in the literature, in particular ISO standards.

Old Units			SI Units		
Name	Symbol	Usage allowed?	Preferred unit	Symbol	Conversion
quantity: length	*symbol: l*	*unit: m*			
foot	ft	no	meter	m	1 ft = 0.3048 m
inch	in	no	millimeter	mm	1 in = 25.4 mm
centimeter	cm	yes	meter	m	1 cm = 0.01 m
micron	μ	no	micrometer	μm	1 μ = 1 μm = 10^{-6} m
millimicron	mμ	no	nanometer	nm	1 mμ = 1 nm = 10^{-9} m
Ångström	Å	no	nanometer	nm	1 Å = 0.1 nm
quantity: lineic number	*symbol: σ*	*unit: m^{-1}*			
per centimeter	cm^{-1}	yes	per meter	m^{-1}	1 cm^{-1} = 100 m^{-1}
mesh	in^{-1}	no	per meter	m^{-1}	1 in^{-1} ≈ 39.4 m^{-1}
quantity: area	*symbol: A*	*unit: m^2*			
square inch	sq.in	no	square millimeter	mm^2	1 sq.in = 645.16 mm^2
quantity: volume	*symbol: V*	*unit: m^3*			
litre	l or L	yes	cubic decimeter	dm^3	1 l = 1 dm^3 = 10^{-3} m^3
decilitre	dl	yes	cubic centimeter	cm^3	1 dl = 100 cm^3
centilitre	cl	yes	cubic centimeter	cm^3	1 cl = 10 cm^3
mililitre	ml, mL	yes	cubic centimeter	cm^3	1 ml = 1 cm^3 = 10^{-6} m^3
microlitre	μl	yes	cubic centimeter	mm^3	1 μl = 1mm^3 = 10^{-9} m^3
quantity: mass	*symbol: m*	*unit: kg*			
ton	t	yes	kilogram	kg	1 ton = 1000 kg
pound	lb	no	kilogram	kg	1 pound = 0.4536 kg
ounce	oz	no	gram	g	1 ounce = 28.3 g

Continued

Old Units			SI Units		
Name	Symbol	Usage allowed?	Preferred unit	Symbol	Conversion
quantity: density symbol: ρ unit: kg m^{-3}					
gram per 100 ml	g/100ml	no	gram per cubic decimeter	g dm^{-3}	1 g/100 ml = 10 g dm^{-3}
percent (w/v)	%	no	gram per cubic decimeter	g dm^{-3}	1 % (w/v) = 10 g dm^{-3}
gram per litre	g/l	yes	gram per cubic decimeter	g dm^{-3}	1 g/l = 1 g dm^{-3}
milligram per litre	mg/l	yes	milligram per cubic decimeter	mg dm^{-3}	1 mg/l = 1 mg dm^{-3}
milligram percent	mg %	no	milligram per cubic decimeter	mg dm^{-3}	1 mg % = 10 mg dm^{-3}
quantity: time symbol: t unit: s					
day	d	yes	second	s	1 day = 86 400 s
hour	h	yes	second	s	1 hour = 3 600 s
minute	min	yes	second	s	1 minute = 60 s
quantity: force symbol: F unit: N = kg m s^{-2}					
kilogramforce (or kg)	kgf	no	newton	N	1kgf ≈ 9.81 N
dyne	dyn	no	newton	N	1 dyn = 10^{-5} N
quantity: pressure symbol: p unit: Pa = kg m^{-1} s^{-2}					
atmosphere	atm	no	pascal megapascal	Pa MPa	1 atm ≈ 10^5 Pa 1 atm ≈ 0.1 MPa
bar	bar	yes	megapascal	MPa	1 bar = 0.1 MPa
millimeter mercury	mm Hg	no	pascal	Pa	1 mm Hg ≈ 133 Pa
torr	torr	no	pascal	Pa	1 torr ≈ 133 Pa
meter water	m H_2O	no	kilopascal	kPa	1 m H_2O ≈ 9.81 kPa
pounds per square inch	p.s.i.	no	kilopascal	kPa	1 p.s.i. ≈ 6.9 kPa
quantity: energy symbol: E unit: J = kg m^2 s^{-2}					
calorie	cal	no	joule	J	1 cal = 4.1868 J
erg	erg	no	nanojoule	nJ	1 erg = 100 nJ = 10^{-7} J
electronvolt	eV	yes	attojoule	aJ	1 eV ≈ 0.16 aJ = 0.16 × 10^{-18} J
kilowatthour	kWh	yes	megajoule	MJ	1 kWh = 3.6 MJ
quantity: power symbol: P unit: W = kg m^2 s^{-3}					
horsepower	hp	no	kilowatt	kW	1 hp ≈ 0.735 kW
voltampere	VA	yes	watt	W	1 VA = 1 W
kilocalorie per hour	kcal/h	no	watt	W	1 kcal/h = 1.163 W

Continued

Old Units			SI Units		
Name	Symbol	Usage allowed?	Preferred unit	Symbol	Conversion
quantity: electric charge		*symbol: Q*	*unit: C = A s*		
faraday	F	no	mol	mol	1 F = 1 mol electrons (or monovalent ions)
			coulomb	C	1 F ≈ 96 500 C
elementary charge	e	yes	coulomb	C	1 e ≈ 0.16 × 10^{-18} C
quantity: lineic conduction		*symbol: κ*	*unit: S m^{-1}*		
reciprocal ohm	mho	no	siemens	S	1 mho = 1 S
millimho per cm	mmho/cm	no	millisiemens per meter	mS m^{-1}	1 mmho/cm = 100 mS m^{-1}
millisiemens per centimeter	mS/cm	yes	millisiemens per meter	mS m^{-1}	1 mS/cm = 100 mS m^{-1}
quantity: temperature		*symbol: T or t*	*unit: K or °C*		
degrees centigrade	°C	yes	kelvin	K	°C ≈ K − 273
degrees Fahrenheit	°F	no	kelvin	K	°F ≈ (9/5)K − 460
quantity: amount of matter		*symbol: n*	*unit: mol*		
gramatom	grat	no	mol	mol	1 grat = 1 mol atoms
grammolecule	gmol	no	mol	mol	1gmol = 1 mol molecules
gramion	gion	no	mol	mol	1 gion = 1 mol ions
equivalent	eq	no	mol	mol	1 eq = 1 mol monovalent ions
quantity: concentration of matter		*symbol: c*	*unit: mol m^{-3}*		
mol per litre (or molarity)	mol/l (M)	yes	mol per cubic decimeter	mol dm^{-3}	1 mol/l = 1 M = 1 mol dm^{-3}
normality	eq/l (N)	no	mol per cubic decimeter	mol dm^{-3}	1 N = 1 mol dm^{-3} of monovalent ions
mol percent	mol %	no	mol per cubic decimeter	mol dm^{-3}	1 mol % = 10 mol dm^{-3}
quantity: matter content		*symbol: n/m*	*unit: mol kg^{-1}*		
milliequiv. per 100 gram	meq/ 100 g	no	centimol per kilogram	cmol kg^{-1}	1 meq/100 g = 1 cmol(+) kg^{-1}
mol per cent	mol %	no	mol per kilogram	mol kg^{-1}	1 mol % = 10 mol kg^{-1}
molality	molal (m)	no	mol per kilogram	mol kg^{-1}	1 molal = 1 mol kg^{-1}

Continued

Old Units			SI Units		
Name	Symbol	Usage allowed?	Preferred unit	Symbol	Conversion
quantity: volume fraction		*symbol: φ*	*unit:* $m^3\ m^{-3}$ *(=1)*		
percent (v/v)	vol %	no	cubic meter per cubic meter	$m^3\ m^{-3}$	1 vol % = 10×10^{-3}
parts per million	ppm	no	cubic meter per cubic meter	$m^3\ m^{-3}$	1 ppm = 1×10^{-6} = = 1 cm^3 per m^3
parts per billion	ppb	no	cubic meter per cubic meter	m3 m^{-3}	1 ppb = 1×10^{-9} = = 1 mm^3 per m^3
millilitre per litre	ml/l	yes	cubic meter per cubic meter	m3 m^{-3}	1 ml/l = 1×10^{-3}
quantity: mass fraction		*symbol: w*	*unit:* $kg\ kg^{-1}$ *(=1)*		
percent (w/w)	w%	no	kilogram per kilogram	$kg\ kg^{-1}$	1 w % = 10×10^{-3}
parts per million	ppm	no	kilogram per kilogram	$kg\ kg^{-1}$	1 ppm = 1×10^{-6} = 1 mg per kg
parts per billion	ppb	no	kilogram per kilogram	$kg\ kg^{-1}$	1 ppb = 1×10^{-9} = 1 µg per kg
milligram percent	mg %	no	kilogram per kilogram	$kg\ kg^{-1}$	1 mg % = 10 mg kg^{-1}
quantity: matter fraction		*symbol: x*	*unit:* $mol\ mol^{-1}$ *(=1)*		
mol percent	mol %	no	mol per mol	$mol\ mol^{-1}$	1 mol % = 10×10^{-3}
parts per million	ppm	no	mol per mol	$mol\ mol^{-1}$	1 ppm = 1×10^{-6}

Prefixes with the SI units

Name	Symbol	Meaning	Name	Symbol	Meaning
exa	E	10^{18}	(deci	d	10^{-1})
peta	P	10^{15}	(centi	c	10^{-2})
tera	T	10^{12}	milli	m	10^{-3}
giga	G	10^{9}	micro	µ	10^{-6}
mega	M	10^{6}	nano	n	10^{-9}
kilo	k	10^{3}	pico	p	10^{-12}
(hecto	h	10^{2})	femto	f	10^{-15}
(deca	d	10^{1})	atto	a	10^{-18}

Remarks:

- Prefixes are written without spacing just before the symbol of the unit. Together they form one new symbol, that is retained as such in composites and exponents.
- The prefixes centi, deci, deca and hecto are permitted, but their use is disregarded.

- It is not allowed to use more than one prefix within one unit symbol; thus not mμm but nm, not μμF but pF, not kpJ but nJ. It is not recommended to use more than one prefix within one expression, so not $mmol/cm^3$ but $kmol/m^3$, not 10 $\mu W/cm^2$ but 0.1 W/m^2.
- The magnitude of the unit in the denominator is no indication of the magnitude of the sample. For example, an amount of matter of 5.5 mmol in 25 cm^3 is not recorded as a matter concentration of 5.5 mmol/25 cm^3, but of 220 mol/m^3; it may be followed by "the sample volume was 25 cm^3".

APPENDIX A-3. STANDARDIZATION OF VOLUMETRIC GLASSWARE

Principle

The unit of volume, the litre, is defined as the volume of 1 kg of water at 4 °C and at normal atmospheric conditions. This definition implies that the standardization of volumetric glassware is only possible by weighing a known quantity of water. Such an approach may, however, introduce several errors because of the buoyancy effect, the expansion of water and the expansion of glass, all of which are temperature-dependent. Thus, computation of the true volume of glassware from weighed quantities of water implies the use of a recalculation factor f, which is listed in Table 1 for several temperatures.

An inscription on volumetric glassware indicates at which temperature the volume is considered to be correct. This temperature is respectively 20, 25 and 27 °C for laboratories in temperate regions, the USA and the tropics. The next paragraphs indicate how the volumes of volumetric flasks, pipettes and burettes should be standardized.

1. VOLUMETRIC FLASKS

Principle

A volumetric flask is filled with water; its true volume is derived from the weight and temperature of its contents.

Procedure

Weigh out a dry and clean volumetric flask. Fill it to the mark with boiled, distilled water that is at room temperature. Measure the weight of the filled flask and record the temperature of the water it contains.

Calculation

The true volume of the flask is:

$$V = (w_1 - w_2) \times f$$

in which:
V = true volume at the chosen standard temperature, in ml;
w_1 = weight of the filled flask, in g;
w_2 = weight of the empty flask, in g;
f = recalculation factor (see Table 1).

2. PIPETTES

Principle

A pipette is filled to the mark with water and its contents is collected in a preweighed bottle. By measuring the weight and temperature of the delivered water, the true volume of the pipette can be calculated.

Procedure

Fill the pipette with boiled water that is at room temperature. Dry the tip with a tissue. Then let water flow away until the lowest part of the meniscus just touches the mark. Whilst doing this, the tip of the pipette must be held against the inner wall of a vessel, at an angle of 45 °. Collect the contents of the pipette in a preweighed weighing bottle, also while holding the tip at an angle of 45 °. Wait for 10 s before removing the pipette's tip from the wall of the weighing bottle. Repeat this procedure four times. Measure the temperature of the water used.

Calculation

The true volume of the pipette is:

$$V = (w_1 - w_2) \times f$$

in which:
V = true volume at the chosen standard temperature, in ml;
w_1 = weight of the filled weighing bottle, in g;
w_2 = weight of the empty weighing bottle, in g;
f = recalculation factor (see Table 1).

3. BURETTES

Principle

The graduations on a burette are checked on at, at least, five different points which should be distributed evenly over the whole scale range. The volume of water corresponding with a prechosen graduation is collected in a bottle. The weight and temperature of this water are then measured; this allows the calculation of the true volume of the burette.

Procedure

Fill the burette with boiled water, that is at room temperature, until the water stands 1 cm above the zero point. Then let, carefully, water run down until the lowest part of the meniscus just touches the zero mark. Remove, by means of a filter paper, all drops that cling to the inner wall of the burette. Remove also any drop that hangs from the tip of the burette, by touching this tip against the wall of a vessel. Run off the contents of the burette into a preweighed bottle until the chosen graduation mark is reached. Collect also the last drop hanging from the tip of the burette. Measure the weight and the temperature of the water used. Thereafter, repeat the above procedure for the other checkpoints. All of these determinations must start at the zero mark of the burette.

Calculation

The true volume of the burette, corresponding with the graduation mark considered, is:

$$V = (w_1 - w_2) \times f$$

in which:
V = true volume at the chosen standard temperature, in ml;
w_1 = weight of the filled weighing bottle, in g;
w_2 = weight of the empty weighing bottle, in g;
f = recalculation factor (see Table 1).

Table 1. Value of the recalculation factor *f* for different temperatures

Standardization for 20 °C		Standardization for 25 °C		Standardization for 27 °C	
temperature (°C)	*f*	temperature (°C)	*f*	temperature (°C)	*f*
14	1.00195	19	1.00272	21	1.0032
15	1.00207	20	1.00291	22	1.0034
16	1.00220	21	1.00311	23	1.0036
17	1.00234	22	1.00332	24	1.0038
18	1.00249	23	1.00355	25	1.0040
19	1.00265	24	1.00378	26	1.0043
20	1.00282	25	1.00402	27	1.0045
21	1.00300	26	1.00428	28	1.0048
22	1.00319	27	1.00454	29	1.0050
23	1.00340	28	1.00481	30	1.0053
24	1.00361	29	1.00509	31	1.0056
25	1.00383	30	1.00538	32	1.0059
26	1.00406	31	1.00568	33	1.0062
27	1.00431	32	1.00599	34	1.0064
28	1.00456	33	1.00631	35	1.0068

APPENDIX A-4. STANDARD SUBSTANCES AND SOLUTIONS; FILTER PAPER

A. Some primary standards and their treatment

With the exception of gravimetric procedures, all laboratory-born results are based on comparison with a standard. So, a correct standard is of utmost importance for reliable analytical results. There are two categories of standards: primary and secondary.

A primary standard should have the following characteristics:

- available in a pure state;
- be unaltered during weighing and storage;
- have high molecular mass;
- be readily soluble;
- react instantaneously and completely.

Some substances used for primary standards are listed in the table below. Each of these has to be dried at optimum temperature just before weighing, according to this (incomplete) table:

Primary standard	Recommended pretreatment	
	Temperature (°C)	Time (h)
Calcium carbonate	105	2
Sodium carbonate	270 ± 10	2
Sodium chloride	≈ 200	24
Sodium oxalate	105	2
Potassium hydrogen phthalate	105	2
Potassium iodate	180	
Arsenic trioxide **(poison!)**	105–110	
Potassium dichromate	150–200	3

Just before drying, any lumps should be cut down so that only fine crystals remain. This is important, because coarse crystals contain much occluded water in their cavities and thus would need a higher drying temperature. The fine crystals should, however, not be pulverized because this enhances deliquescence (= attraction of water) and/or efflorescence (= weathering) later on. Besides, the heating oven should be ventilated well, since this decreases both the time and the temperature of drying. After the recommended heating procedure, the primary standard substance should be cooled in a desiccator containing preferably magnesium perchlorate as a dessicant. The dried substance should be weighed as soon as it has reached the ambient temperature.

Primary standard substances should be stored in closed weighing bottles which are placed in a desiccator. The desiccator may contain a desiccant like silicagel (which should be blue). In this way, a primary standard may be stored for weeks or months without appreciable uptake of water. It is advisable, however, to repeat the drying procedure now and then.

As a rule, hydrated substances are not suitable for use as a primary standard, because of difficulties in keeping the amount of crystal water constant.

Some substances are not allowed to be dried, even though they do not contain crystal water. This holds e.g. for boric acid, H_3BO_3, because drying would convert it to metaboric acid, HBO_2.

B. Secondary standard substances

A secondary standard is a substance which may be used for standardizations after having been compared against some primary standard. Many hydrated reagents fall into this category. Two of them are of practical importance: oxalic acid and borax.

Oxalic acid, $(COOH)_2 \cdot 2H_2O$, may be stored over a saturated solution of NaBr. It may be more convenient, however, to check the content by titrating with standardized $KMnO_4$. In either case it seems wise to purchase analytical grade oxalic acid.

Borax, $Na_2B_4O_7 \cdot 10H_2O$, has the advantage of a high molecular mass in comparison to oxalic acid. Borax may be stored over a saturated solution of NaCl or NaBr. In the authors' laboratory, a sample of borax decahydrate (stored in the bottle in which it had been supplied) lost 25% of its crystal water in a period of 2 years! They recrystallize borax instead of storing it over a deliquescing salt (see App. A-26 and A-27).

Hydrated substances should never be pulverized, because this accelerates the process of efflorescence greatly.

C. Preparation of standard solutions

Reagents: The purity of reagents should be chosen in accordance with the intended applications. It is common practice to use the "analytical reagent" quality, which contains contaminations at mg/kg level. For routine analysis of macro elements, however, the "chemically pure" grade might suffice. For the determination of micro elements, on the other hand, it might be necessary to use high-purity chemicals, which are specified for different techniques ("spectroscopically pure", "chromatographically pure"). Since such reagents are very expensive, the need for their application should be established first.

Water: The water used for laboratory work has to meet several requirements concerning its purity, among which pH and conductivity are the main ones. The pH may vary between 5.0 and 7.5, while the electrical conductivity should be lower than 100 mS m^{-1}.

Two ways of purification are well-known and practised generally: **distillation** and **deionisation** (the latter is also called demineralisation). Each technique is appropriate for normal analytical work. However, after single distillation (from hard glass apparatus) traces of metal ions are still present; deionisation, on the other hand, may leave traces of organic substances in the water.

For special purposes, e.g. determination of very low contents of trace elements, the two techniques may be applied both, thus producing "demi-dest" water. In some cases, e.g. for the titrimetric determination of low levels of carbonate, it is necessary to use CO_2-free water. This is simply obtained by boiling for 4 minutes; during cooling the vessel should be stoppered by a guard tube containing a CO_2 absorbent, but in practice a watch glass has proved to be sufficient.

Solutions: All standard substances are only applied in solution, and are accordingly called primary standard solutions and secondary standard solutions. The preparation of such solutions (preferably in 0.1 M acid) is usually straightforward: weighing, dissolving and making up to volume. Of course, the substance must be weighed very precisely and prepared in a volumetric flask. In this way, solutions of rather high concentrations (about 0.1–1 M) are prepared; these are called **stock solutions**. For several substances there are stock solutions commercially available in sealed ampoules. This may be convenient, e.g. when the standard substance does not dissolve readily.

Stock solutions are preferably stored cool and dark, either in hard glass or in polythene vessels, flushed in advance with acid, and may thus be stable for months. When stock solutions are stored for more than a year, they should be checked before using; it is known, for example, that storage in a polythene flask may lead to 1–1.5% loss of water by evaporation.

From those stock solutions all other solutions are made by dilution, these are called **standard solutions**. Standard solutions may deteriorate rather rapidly due to adsorption, irradiation or the like. This holds in particular for very diluted solutions. It is advisable to check the rate of decomposition, and to prepare fresh standard solutions at an appropriate frequency, if necessary every day.

D. Filter paper

Filter paper is classified from coarse to fine according to its porosity. The following table gives this classification for some well-known brands of quantitative papers, i.e. HCl-HF treated papers (formerly called "ash-free").

porosity	Macherey & Nagel	Schleicher & Schüll	Whatman	
			normal	hardened*
quantitative paper				
very fine	MN 640 dc	S/S 589/3	42	542
fine	MN 640 d	S/S 589/5		
medium fine	MN 640 m	S/S 589/2	40	540
coarse	MN 640 w	S/S 589/1	41	541
phase separating paper				
medium	MN 616 WA	S/S 597 hy	1PS	

* especially for filtration with Büchner funnels.

REFERENCES

T. Yoshimori. Drying and weighing of standard reference materials for titrimetric analysis and the status of the Faraday constant as an international standard. Talanta *22* (1975) 827–825.

APPENDIX A-5. FREQUENTLY USED INDICATORS AND THEIR PREPARATION

Type	Name	Transition range (pH)	Colour change	
			Acid	*Base*
Acid-base	Bromocresol green	3.8–5.4	yellow	blue-green
	Methyl red	4.2–6.3	red	yellow
	Mixed indicator (b.g. & m.r.)	5.1	red	green
	Bromothymol blue	6.0–7.6	yellow	blue
	Cresol red	7.2–8.8	yellow	purple
	Phenolphtalein	8.2–10.0	colourless	red
			Metal complex	*Free*
Metal ion	Eriochrome Black T	7–12	wine red	blue
	Murexide	9–11	yellow/red	purple

Recommended preparations

Bromocresol green: Suspend 0.15 g in 100 ml ethanol 96% (v/v). Add dropwise NaOH 0.1 M until the red colour turns into dark red (about 1.5 ml). Dissolve by stirring. If the indicator is available as its sodium salt, it is sufficient to dissolve 0.15 g in 100 ml of ethanol 96% (v/v).

Methyl red: Dissolve 0.1 g in 100 ml of ethanol 96% (v/v).

Mixed indicator: Mix equal volumes of the above solutions.

Bromothymol blue: Dissolve 0.1 g in 100 ml of water to which 1.6 ml 0.1 M NaOH have been added.

Cresol red: Dissolve 0.1 g in 100 ml ethanol 20% (v/v).

Phenolphtalein: Dissolve 0.1 g in 100 ml of ethanol 90% (v/v).

Eriochrome Black T: Dissolve 0.5 g in 100 ml of ethanol 96% (v/v). Add 4.5 g of hydroxylamine-HCl and stir on a magnetic stirrer for at least 30 min. The hydroxylamine-HCl will dissolve partially. Transfer the solution and as much as possible of the remaining hydroxylamine-HCl crystals to an amber coloured storage bottle. The excess of crystals will conserve the indicator for at least 6 months.

Murexide: Suspend 50 mg in 5 ml water to obtain a saturated solution. Filter. Prepare fresh daily.

APPENDIX A-6. STORAGE AND TREATMENT OF PERCHLORIC ACID

Perchloric acid is a hazardous chemical. Nevertheless, it can be used with safety if the user knows its properties and applies the necessary precautions.

Properties

1. Under normal storage conditions perchloric acid is stable at concentrations of less than 85% (w/w). The acid usually sold, has concentrations of 60 to 72% (w/w) (d = 1.6 g/cm^3).
2. The azeotropic mixture with water contains 72.5% (w/w) of perchloric acid and boils at 203 °C (at 10^5 Pa pressure). This means that the evaporation of an aqueous solution of the acid can never produce an acid of dangerously high concentration. In presence of salts of metals the solution should never be evaporated to dryness as this may cause explosions.
3. Perchloric acid vapour and inflammable gases form explosive mixtures.
4. Hot 60–72% perchloric acid is a very strongly oxidizing agent, which attacks any form of organic matter.
5. Cold and diluted perchloric acid loses its oxidizing properties.
6. Perchloric acid solutions are normally colourless.

Use and storage

1. Store solutions of perchloric acid in glass–stoppered flasks standing in a dish made of porcelain.
2. Store flasks containing the acid *separately* from flasks containing reducing solutions such as alcohol, glycerol, hypophosphites.
3. After spillage of the acid, flush thoroughly with water.
4. If perchloric acid has become coloured, dilute the acid with water and discard the solution.
5. If perchloric acid is used for the digestion of organic matter, an excess of nitric acid should be present to attack the organic matter first. The nitric acid will moderate the reaction by oxidizing the easily oxidizable components at a lower temperature. Some of the more stable substances may be left and will react later on with perchloric acid at a higher temperature. Oil or fat containing products such as oilpalm seeds, peanuts etc. may cause troubles.
6. During a digestion with perchloric acid safety goggles should be worn.
7. Digestions have to be executed in suitable fume hoods. The construction of the chimney should allow periodical washings with water. Only a ventilator of the centrifugal type (motor outside the stream of gases) can be used. The vapours should neither come into contact with any metal parts nor with easily oxidisable material.
8. To prevent condensation of the acid, an efficient draught is required. After completion of the digestion the fan should be left on for half an hour.

REFERENCES

Analytical Methods Committee. Notes on perchloric acid and its handling in analytical work. Analyst *84* (1959) 214–216.

APPENDIX A-7. PROPERTIES AND USAGE OF PLATINUM WARE

1. Properties

- Platinum melts at 1770 °C and becomes soft when approaching that temperature.
- Platinum can stand the action of most salts.
- Platinum is hardened with an alloy that contains iridium and/or rhodium. This alloy does not resist the action of acids as good as pure platinum does. Moreover, some of the iridium will volatilize during heating, which will result in loss of weight.
- Hammered-out crucibles are distorted less easily than spun ones.

2. Quality requirement

- Platinum should not show a change in colour when heated.
- The crucible should be free of iron after treatment with an acid.
- Four hours of heating at 1100 °C should not result in a loss of weight larger than 0.2 mg/hour.
- The content of iridium should be 5 % at the most.

3. Cleaning

Cleaning should be done, in this sequence, by the following procedure:

- Remove organic matter with chromic acid.
- Dissolve carbonates and metal oxides by boiling with either HCl or HNO_3. However, do *not* use a mixture of these two acids.
- Remove silicates by fusion over a flame in the presence of Na_2CO_3 or $Na_2B_4O_7$.
- Remove metals and metal oxides by fusion over a flame in the presence of sodium hydrogen sulphate or potassium hydrogen sulphate.
- Finally, heat the crucible carefully over a flame.

4. Care and uses

When working with platinum vessels, do:

- never use mixtures of HCl or HNO_3;
- never use solutions containing free chlorine;
- never create long contacts with acid solutions of ferric chloride;
- never melt sodium- and potassium hydroxides;
- never heat with compounds that contain Pb, Sn, Bi, As, Sb or Zn, since easily reducible metal oxides form alloys with platinum;
- never heat with sulphides;

- take care in heating compounds that contain high concentrations of phosphorus, since reduction by hot carbon from the flame forms platinum phosphide;
- never fuse with mixtures that might produce halogens in the presence of oxidizing compounds such as permanganate or chromate;
- never fuse over a luminous flame as this may produce carbides of platinum;
- never put platinum crucibles on a metal base while heating in a furnace, but use asbestos instead;
- never use C, CN, CNS, halogenides and arsenites as they form complexes with platinum.

Moreover, *take care* to:

- use only tongs with platinum tips;
- put always a lid on the crucible when performing an alkaline melt with Na_2CO_3, in order to exclude the action of oxygen from the atmosphere;
- use sodium- instead of potassium compounds, because the former are less corroding. Generally, the corrosion increases strongly at temperatures above 700 °C, whereas this temperature should be lower than 500–600 °C in case of KOH, $Ba(OH)_2$, peroxides and cyanides. Up to 800 °C is permitted for carbonate and neutral salts;
- cool the vessel carefully after heating, so as to prevent the formation of minute cracks.

APPENDIX A-8. GRAVIMETRIC DETERMINATION OF Ca (AS CARBONATE VIA OXALATE PRECIPITATION)

Principle

Calcium is precipitated as oxalate and finally weighed as carbonate after heating in a furnace at a temperature of 475–525 °C. Large quantities of metals such as Cu, Pb and Zn should be absent as they yield slightly soluble oxalates. Coprecipitation of Mg may be diminished by not boiling the solution and by shortening the time period before filtration. Furthermore, a very large excess of oxalate should be avoided to prevent the formation of a Mg-oxalate complex which is much more soluble.

Reagents

Hydrochloric acid, c(HCl) = 6 mol/l. Add 50 ml of concentrated hydrochloric acid (36%) to about 40 ml water and make up to 100 ml.

Methyl red indicator. See Appendix A-5.

Ammonia, $c(NH_3)$ = 6.7 mol/l. Dilute 250 ml of concentrated aqueous ammonia (25%) with water to 500 ml.

Ammonium oxalate solution, 4%. Dissolve 4 g of ammonium oxalate in water and make up to 100 ml.

Diluted ammonium oxalate solution, 0.1–0.2%. Dilute about 3–5 ml of the 4% ammonium oxalate solution with water to 100 ml.

Procedure

Transfer a sample containing less than 0.2 g of Ca to a 400-ml beaker, add 10 ml of water and 15 ml of 6 M hydrochloric acid. Heat gently and boil for some time to expel the carbon dioxide. Dilute to 200 ml, add 2 drops of the indicator, heat to boiling and add – very slowly – to the still hot solution 50 ml of a warm solution of 4% ammonium oxalate. Then add dropwise 6.7 M ammonia until the solution is neutral or weakly alkaline. Leave the solution without further heating for at least 1 hour. Then test it for complete precipitation.

Transfer the solution (by decanting) to an ashfree filter. Wash the filter at least five times with diluted ammonium oxalate solution. Then transfer the filter to a preweighed crucible, put it in an oven at 105 °C for 1 hour and next in a furnace at 500–525 °C for 2 hours. Cool the crucible, first in the air for about 15 min and then in a desiccator for 30 min. Then weigh again. Next, put the crucible again in the furnace at 500–525 °C, but now for 1/2 hour. Cool in the same way and weigh. Repeat the last procedure until constant weight.

Calculation

The weight percentage of calcium is:

$$\text{Ca (in \%)} = 40.04 \times \frac{w_2}{w_1}$$

in which:
w_1 = initial weight of the sample, in g;
w_2 = weight of $CaCO_3$ remaining after heating in the furnace, in g.

Remark:

- Take care not to raise the furnace temperature above 525 °C in order to avoid decomposition of the calcium carbonate.

APPENDIX A-9. STANDARDIZATION OF 0.1 M HCl OR 0.05 M H_2SO_4 WITH SODIUM CARBONATE

Principle

The acid is standardized by titration of sodium carbonate with methyl red-bromocresol green as an indicator.

Reagents

Sodium carbonate. Na_2CO_3, pretreated according to App. A-4.
Mixed indicator. Methyl red-bromocresol green (see App. A-5).

Procedure

Weigh out precisely about 200 mg of sodium carbonate and transfer it to a 300-ml erlenmeyer flask. Dissolve in about 75 ml of distilled water and add 3 drops of the indicator. Titrate with the acid until the colour has changed from green via grey to rose. Then boil until the CO_2 is driven out of the solution, cool and titrate further until the colour has become rose again.

Calculation

The titer of the acid is:

$$t = 0.01887 \times \frac{w}{V} \text{ for HCl}$$

or

$$t = 0.009439 \times \frac{w}{V} \text{ for } H_2SO_4$$

in which:
t = titer (concentration) of the acid, in mol/l;
w = weight of sodium carbonate, in mg;
V = volume of acid added, in ml.

Remarks:

- The colour change before the boiling stage should be reached 0.2–0.3 ml before the equivalence point. After boiling, the consumption of acid should not exceed 0.3 ml, otherwise the determination should be repeated.
- Dried sodium carbonate is very hygroscopic.

APPENDIX A-10. STANDARDIZATION OF 0.1 M NaOH OR KOH WITH OXALIC ACID

Principle

The base is standardized by titration of oxalic acid with methyl red as an indicator.

Reagents

Oxalic acid. $(COOH)_2 \cdot 2H_2O$, pretreated according to App. A-4.

Calcium chloride. Dissolve 200 g of $CaCl_2 \cdot 6H_2O$ in 1 litre of water; neutralize with respect to methyl red.

Indicator solution. Methyl red (see App. A-5).

Procedure

Weigh out precisely about 250 mg of oxalic acid and transfer it to a 300-ml erlenmeyer flask. Dissolve the acid in about 100 ml of distilled water (free from CO_2), add 3 drops of indicator solution and titrate with the base until the colour of the indicator has changed to yellow. Then add 10 ml of the calcium chloride solution and proceed with the titration until the colour has once again changed to yellow.

Calculation

The titer of the base is:

$$t = 0.01586 \times \frac{w}{V}$$

in which:

t = titer (concentration) of the base, in mol/l;

w = weight of oxalic acid, in mg;

V = volume of base, in ml.

Remarks:

- Sodium oxalate reacts alkaline with respect to the indicator. Therefore, calcium chloride is added which results in a precipitation of calcium oxalate and the consequent liberation of HCl; this HCl is neutralized in the final stage of titration.
- To prevent coprecipitation of oxalic acid, the solution of calcium chloride is added at the end of the titration.
- Small amounts (200–500 ml) of CO_2-free water can be prepared by boiling during 4 minutes in erlenmeyer flasks (*not* in beakers). It is allowable to cool while only a watch glass covers the erlenmeyer flask.

APPENDIX A-11. STANDARDIZATION OF 0.1 M HCl OR 0.05 M H_2SO_4 WITH POTASSIUM IODATE

Principle

A solution containing a known quantity of KIO_3, together with some KI and $Na_2S_2O_3$, is titrated with acid. During this process I_2 is formed, which is reduced to I^- by $Na_2S_2O_3$. After the equivalence point (i.e., when all KIO_3 is consumed), any excess of acid will lower the pH which will be revealed by an acid-base indicator.

Reagents

Potassium iodate. KIO_3, pretreated according to App. A-4.
Sodium thiosulphate. $Na_2S_2O_3$.
Potassium iodide. KI.
Mixed indicator solution. Methyl red-bromocresol green (see App. A-5).

Procedure

Weigh out precisely about 125 mg of potassium iodate; weigh out also approximately 1.0 g of potassium iodide and approximately 1.2 g of sodium thiosulphate. Transfer these substances to a 300-ml erlenmeyer flask and dissolve them in about 100 ml of water. Then add 3 drops of the mixed indicator solution and titrate with the acid until the colour of the indicator has changed from green, via grey, to rose. The rose colour should persist for 3 minutes.

Perform also a blank determination (i.e., without the KIO_3).

Calculation

The titer of the acid is:

$$t = 0.02804 \times \frac{w}{a - b} \quad \text{for HCl}$$

or

$$t = 0.01402 \times \frac{w}{a - b} \quad \text{for } H_2SO_4$$

in which
t = titer (concentration) of the acid, in mol/l;
w = weight of KIO_3, in mg;
a = volume of acid used for the analyte solution, in ml;
b = volume of acid used for the blank solution, in ml.

APPENDIX A-12. STANDARDIZATION OF 0.02 M $KMnO_4$ WITH SODIUM OXALATE

Principle

A $KMnO_4$ solution is standardized by oxidimetric titration of sodium oxalate.

Reagents

Sodium oxalate. $(COONa)_2$, pretreated according to App. A-4.

Sulphuric acid, $c(H_2SO_4) = 2$ mol/l. Add carefully, while swirling, 56 ml of concentrated sulphuric acid (96%) to about 400 ml water. Allow to cool and make up to 500 ml.

Potassium permanganate solution, $c(KMnO_4) = 0.02$ mol/l. Dissolve 3.2 g $KMnO_4$ in about 800 ml of hot distilled water and then boil for about 5 min. Cool, filter over a 0.45-μm membrane filter or a very fine glass frit and make up to 1 litre with water.

Procedure

Weigh out precisely about 275 mg of sodium oxalate and transfer it to a beaker of 600 ml. Dissolve the substance, while stirring, in a mixture of 100 ml of 2 M sulphuric acid and 100 ml of water at about 27 °C.

While stirring, add with a burette about 35 ml of the permanganate solution and wait until the solution has become colourless. Heat to 55–60 °C and then titrate dropwise, while stirring, until a rose colour persists after a waiting period of 30 seconds.

Perform a blank determination with a mixture of 100 ml of 2 M sulphuric acid and about 140 ml of water.

Calculation

The titer of the permanganate solution is:

$$t = 0.002987 \times \frac{w}{(a - b)}$$

in which

t = titer (concentration) of the permanganate solution, in mol/l;
w = weight of the sodium oxalate, in mg;
a = volume of $KMnO_4$ added to the analyte solution, in ml;
b = volume of $KMnO_4$ added to the blank solution, in ml.

Remark:

- $KMnO_4$ can also be standardized with EDTA (see App. A-21).

APPENDIX A-13. STANDARDIZATION OF 0.1 M $AgNO_3$ WITH SODIUM CHLORIDE (ACCORDING TO MOHR)

Principle

A known amount of NaCl is titrated with an $AgNO_3$ solution, so that silver chloride precipitates. After the equivalence point, any excess of $AgNO_3$ forms a red precipitate of silver chromate with the indicator.

Reagents

Silver nitrate solution, $c(AgNO_3) = 0.1$ mol/l. Dissolve 17.0 g of $AgNO_3$ in 1 litre of water.

Potassium chromate solution. Dissolve 5 g of K_2CrO_4 in 100 ml of water.

Sodium chloride. NaCl, pretreated according to App. A-4.

Calcium carbonate. $CaCO_3$ (analytical grade).

Procedure

First perform a blank determination in a 250-ml beaker that contains 140 ml of water, 4 ml of potassium chromate solution and 0.5 g of $CaCO_3$. Titrate with the silver nitrate solution until the suspension shows a weak, but distinct, red colour which persists even with energetic stirring. Keep this suspension for future comparison.

Then weigh out precisely about 200 mg of sodium chloride, transfer it to a 250-ml beaker, and add 100 ml of water and 4 ml of potassium chromate solution. Titrate carefully, while stirring, until the red colour which appears with every drop of $AgNO_3$ fades away slowly. Then titrate dropwise until the solution shows the same shade of red as the blank.

Calculation

The titer of the silver nitrate solution is:

$$t = 0.01711 \times \frac{w}{(a - b)}$$

in which

t = titer (concentration) of the silver nitrate solution, in mol/l;
w = weight of NaCl, in mg;
a = volume of $AgNO_3$ added to the analyte solution, in ml;
b = volume of $AgNO_3$ added to the blank solution, in ml.

Remarks:

- Because the estimation of the end point is rather subjective, it is necessary to perform a blank.
- The Mohr method is used here only for standardizing the silver nitrate solution, since application to plant material extracts has shown to give interferences.

APPENDIX A-14. STANDARDIZATION OF 0.05 M EDTA WITH CALCIUM CARBONATE

Principle

An EDTA solution is standardized by titration of $CaCO_3$ at pH 10 with Eriochrome Black T as an indicator.

Reagents

Hydrochloric acid, c(HCl) = 1 mol/l. Add 83 ml of concentrated hydrochloric acid (36%) to about 400 ml water and make up to 1 litre.

Buffer solution, pH 10. Dissolve 54 g of ammonium chloride, NH_4Cl, and 2 g of magnesium-EDTA, Mg-EDTA, in 350 ml of concentrated aqueous ammonia (25%) and dilute with water to 1000 ml.

EDTA solution, 0.05 mol/l. Dissolve 18.6 g of disodium ethylenediamine tetra acetic acid, $Na_2EDTA \cdot 2H_2O$, in 1 litre of water.

Indicator solution. Eriochrome Black T solution, prepared according to App. A-5.

Procedure

Weigh out precisely about 150 mg of $CaCO_3$ into a 300-ml erlenmeyer flask. Dissolve it in a small excess of 1 M HCl (about 3 ml). Dilute to about 100 ml and boil for some minutes. Then add to the still hot solution 10 ml of the buffer solution and 0.4 ml of the indicator solution. Titrate from red to a blue end point with the EDTA solution.

Perform also a blank determination.

Calculation

The titer of the EDTA solution is:

$$t = 0.009991 \times \frac{w}{(a - b)}$$

in which

t = titer (concentration) of the EDTA solution, in mol/l;
w = weight of $CaCO_3$, in mg;
a = volume of EDTA solution used for the analyte, in ml;
b = volume of EDTA solution used for the blank, in ml.

Remarks:

- If, by accident, too much HCl has been added, the solution can be neutralized first with 1 M ammonia.
- Store the standardized EDTA solution in polythene bottles.
- The Na_2EDTA should be of analytical grade, since lower grades may contain other complexing agents which can give erroneous results (e.g. in the determination of Zn).

- The addition of Mg-EDTA to the buffer solution provides a sharper colour change at the end point when using Eriochrome Black T.

APPENDIX A-15. STANDARDIZATION OF $KAl(SO_4)_2 \cdot xH_2O$ WITH EDTA

Purpose

Since $KAl(SO_4)_2 \cdot 12H_2O$ may lose crystal water on standing, the aluminium content of this salt should be determined before it is used as a standard.

Principle

To a known amount of $KAl(SO_4)_2 \cdot xH_2O$ an excess of EDTA solution is added. The Al-EDTA complex is formed at pH 6.0–7.0 after boiling the mixture. The excess of EDTA is back-titrated with $ZnSO_4$ solution at pH 7.2–7.8 with Eriochrome Black T as an indicator.

Reagents

EDTA solution, 0.05 mol/l. Standardized according to App. A-14.

Ammonia solution, $c(NH_3)$ = 4 mol/l. Dilute 75 ml of concentrated aqueous ammonia (25%) with water to 250 ml.

Buffer solution pH 10. Dissolve 54 g of ammonium chloride, NH_4Cl, in 350 ml of concentrated aqueous ammonia (25%) and dilute with water to 1000 ml.

p-Nitrophenol. Dissolve 0.1 g in 100 ml water.

Indicator solution. Eriochrome Black T, prepared according to App. A-5.

Zinc solution, 0.05 mol/l. Dissolve 14.37 g of $ZnSO_4$ in 1 litre of water.

Procedure

Pipette 35.00 ml of 0.05 M EDTA solution into a 300-ml erlenmeyer flask. Dilute to about 100 ml and add 3 drops of p-nitrophenol. Add 1.3 ml of 4 M ammonia solution. The solution should now show a distinct yellow colour. Add 0.2 ml of the Eriochrome Black T indicator, whereupon the colour of the solution will become blue-green. Titrate with the 0.05 M zinc solution from blue-green to a wine-red end point (= B ml).

Weigh out precisely about 0.71 g of $KAl(SO_4)_2 \cdot xH_2O$ in a 300-ml erlenmeyer flask. Dissolve it in about 100 ml of water. Add 3 drops of p-nitrophenol and 1.3 ml of 4 M ammonia solution. The solution will show a distinct yellow colour. Add 35.00 ml of 0.05 M EDTA solution and boil during 5 min. Cool to room temperature. Add dropwise 4 M ammonia solution (1–3 drops) until the solution shows again a distinct yellow colour (pH 7.2–7.8). Then add 0.2 ml Eriochrome Black T, whereupon the colour of the solution should change to blue-green. (When the solution is dirty blue, add 1–3 drops of 4 M ammonia solution extra until a blue-green colour is formed). Titrate with the 0.05 M zinc solution from blue-green to a wine-red end point (= A ml).

Calculation

The actual molecular weight (m_x) of the used potassium aluminium sulphate is:

$$m_x = 0.2857 \times \frac{w \times B}{t \times (B - A)}$$

in which
w = weight of $KAl(SO_4)_2 \cdot xH_2O$, in mg;
t = titer of the EDTA solution, in mol/l;
B = volume of zinc solution used for titration of 35.00 ml EDTA, in ml;
A = volume of zinc solution used for titration of excess EDTA, in ml.

Remarks:
- If x = 12, then m = 474.39; when preparing standard solutions of aluminium, using potassium aluminium sulphate with deviating m, one has to weigh out m_x/m times the prescribed amount.
- Enough ammonia should be present to neutralize the hydrogen ions formed during the complexation of aluminium. However, a large excess of ammonia should be avoided to prevent the dissociation of Al-EDTA. Otherwise, back-titration with Zn would become impossible because of the irreversible formation of an Al-Eriochrome Black T complex.
- During boiling the excess of ammonia will volatilize. (After cooling the colour is normally a faint yellow). Therefore, 1–5 drops of 4 M ammonia solution are added to obtain a pH of 7.2–7.8 which yields a distinct yellow colour. When the pH is still too low, the solution will show a dirty blue colour after addition of Eriochrome Black T.

APPENDIX A-16. STANDARDIZATION OF $Cd(NO_3)_2 \cdot xH_2O$ WITH EDTA

Purpose

Since $Cd(NO_3)_2 \cdot 4H_2O$ may absorb water on standing, the cadmium content of this salt should be determined before it is used as a standard.

Principle

A solution of cadmium nitrate is titrated with EDTA at pH 10 with Eriochrome Black T as an indicator.

Reagents

Buffer solution pH 10. Dissolve 54 g of ammonium chloride, NH_4Cl, in 350 ml of concentrated aqueous ammonia (25%) and dilute with water to 1000 ml. (see remark)

EDTA solution, 0.05 mol/l. Standardized according to App. A-14.

Indicator solution. Eriochrome Black T, prepared according to App. A-5.

Procedure

Weigh out precisely about 0.46 g of $Cd(NO_3)_2 \cdot xH_2O$ in a 300-ml erlenmeyer flask. Dissolve it in about 100 ml of water. Add 10 ml of the buffer solution and 0.4 ml of the indicator solution. Titrate from red to a blue end point with standardized 0.05 M EDTA solution.

Perform also a blank determination.

Calculation

The actual molecular weight (m_x) of the used cadmium nitrate is:

$$m_x = \frac{w}{t \times (a - b)}$$

in which

w = weight of $Cd(NO_3)_2 \cdot xH_2O$, in mg;

t = titer of the EDTA solution, in mol/l;

a = volume of EDTA solution used for the analyte, in ml;

b = volume of EDTA solution used for the blank, in ml.

Remarks:

- If x = 4, then m = 308.48; when preparing standard solutions of cadmium, using cadmium nitrate with deviating m, one has to weigh out m_x/m times the prescribed amount.
- The buffer solution used for the standardization of EDTA (App. A-14) may be used as well.
- Cadmium is a very poisonous element. Work in a fume hood and make use of pipette fillers.

APPENDIX A-17. STANDARDIZATION OF $CoCl_2 \cdot xH_2O$ WITH EDTA

Purpose

Since $CoCl_2 \cdot 6H_2O$ may absorb water on standing, the cobalt content of this salt should be determined before it is used as a standard.

Principle

A solution of cobalt chloride is titrated with EDTA at pH 10 in the presence of tartrate with murexide as an indicator.

Reagents

Tartrate solution pH 10. Dissolve 3.75 g of tartaric acid, $C_4H_6O_6$, in 35 ml of 4 M ammonia and dilute with water to 250 ml.

EDTA solution, 0.05 mol/l. Standardized according to App. A-14.

Murexide indicator. Suspend 50 mg of murexide in 5 ml of water to get a saturated solution. Filter. Prepare a fresh solution each day.

Procedure

Weigh out precisely about 0.36 g of $CoCl_2 \cdot xH_2O$ in a 300-ml erlenmeyer flask. Dissolve it in about 100 ml of water. Heat to 70–80 °C and add 20 ml of the tartrate solution and 5 drops of murexide indicator. Titrate from orange-brown to a violet end point with standardized 0.05 M EDTA solution.

Calculation

The actual molecular weight (m_x) of the used cobalt chloride is:

$$m_x = \frac{w}{t \times a}$$

in which

w = weight of $CoCl_2 \cdot xH_2O$, in mg;

t = titer of the EDTA solution, in mol/l;

a = volume of EDTA solution used, in ml.

Remarks:

- If $x = 6$, then $m = 237.94$; when preparing standard solutions of cobalt, using cobalt chloride with deviating m, one has to weigh out m_x/m times the prescribed amount.
- Cobalt is a poisonous element. Work in a fume hood and make use of pipette fillers.
- An aqueous solution of $CoCl_2$ has a pink-red colour. After addition of murexide the colour becomes orange-brown. The Co-EDTA complex also shows an intense pink-red colour; therefore enough murexide should be added so as to get a clearly visible colour change from orange-brown to violet.

APPENDIX A-18. STANDARDIZATION OF $CuSO_4 \cdot xH_2O$ WITH EDTA

Purpose

Since $CuSO_4 \cdot 5H_2O$ may lose crystal water on standing, the copper content of this salt should be determined before it is used as a standard.

Principle

A solution of copper sulphate is titrated with EDTA at pH 10 with murexide as an indicator.

Reagents

Ammonia solution, $c(NH_3)$ = 4 mol/l. Dilute 75 ml of concentrated aqueous ammonia (25%) with water to 250 ml.

EDTA solution, 0.05 mol/l. Standardized according to App. A-14.

Murexide indicator. Suspend 50 mg of murexide in 5 ml of water to obtain a saturated solution. Filter. Prepare a fresh solution each day.

Procedure

Weigh out precisely about 0.37 g of $CuSO_4 \cdot xH_2O$ in a 300-ml erlenmeyer flask. Dissolve it in about 100 ml of water. Add, dropwise, about 2 ml of 4 M ammonia solution. First, a precipitate is formed. Continue to add ammonia solution until this precipitate has just redissolved. The colour of the solution is now dark blue (a slight turbidity does not influence the result). Avoid a large excess of ammonia. Add 4 drops of the murexide indicator and titrate with standardized 0.05 M EDTA solution until the colour changes from green to deep blue.

Calculation

The actual molecular weight (m_x) of the used copper sulphate is:

$$m_x = \frac{w}{t \times a}$$

in which

w = weight of $CuSO_4 \cdot xH_2O$, in mg;

t = titer of the EDTA solution, in mol/l;

a = volume of EDTA solution used, in ml.

Remarks:

- If x = 5, then m = 249.69; when preparing standard solutions of copper, using copper sulphate with deviating m, one has to weigh out m_x/m times the prescribed amount.
- A high concentration of ammonia with respect to the copper concentration results in the formation of a copper-tetrammine complex which will partially decompose the copper-murexide complex. In that case the colour change becomes difficult to observe.
- Copper-murexide has a yellow colour, but in the beginning of the titration the colour is dirty-blue (due to a mixture of copper-ammonia and copper-murexide complexes). Later on, the colour changes to green (due to a mixture of copper-murexide and copper-EDTA complexes). At the endpoint the solution becomes deep blue caused by the mixed colours of free murexide and copper-EDTA.

APPENDIX A-19. STANDARDIZATION OF $(NH_4)_2Fe(SO_4)_2 \cdot xH_2O$ WITH EDTA

Purpose

Since $(NH_4)_2Fe(SO_4)_2 \cdot 6H_2O$ may lose crystal water on standing, the iron content of this salt should be determined before it is used as a standard.

Principle

A solution of ammonium iron(II) sulphate is titrated with EDTA at pH 2.5 with sulfosalicylic acid as an indicator.

Reagents

Nitric acid, $c(HNO_3)$ = 0.1 mol/l. Dilute 7 ml of concentrated nitric acid (65%) with water to 1 litre.

Hydrogen peroxide, 30%. Analytical grade.

Acetate buffer solution. Dissolve 65 g of sodium acetate, $CH_3COONa \cdot 3H_2O$,in 100 ml of 2 M HCl and dilute with water to 250 ml.

EDTA solution, 0.05 mol/l. Standardized according to App. A-14.

Sulfosalicylic acid indicator. Dissolve 2 g of sulfosalicylic acid, $C_7H_6O_6S$, in 100 ml of ethanol 96%.

Procedure

Weigh out precisely about 0.58 g of $(NH_4)_2Fe(SO_4)_2 \cdot xH_2O$ in a 600-ml beaker. Dissolve it in 100 ml of 0.1 M HNO_3. Heat to boiling and add 5–7 drops of hydrogen peroxide to oxidize Fe^{2+} to Fe^{3+}. Boil for a few minutes to expel the excess of H_2O_2 and cool off. Then raise the pH to 2.5 by adding acetate buffer solution (about 10 ml). Add 2 ml of the sulfosalicylic acid indicator and titrate with standardized 0.05 M EDTA solution until the colour changes from red to yellow.

Calculation

The actual molecular weight (m_x) of the used ammonium iron(II) sulphate is:

$$m_x = \frac{w}{t \times a}$$

in which

w = weight of $(NH_4)_2Fe(SO_4)_2 \cdot xH_2O$, in mg;

t = titer of the EDTA solution, in mol/l;

a = volume of EDTA solution used, in ml.

Remarks:

- If x = 6, then m = 392.14; when preparing standard solutions of iron, using ammonium iron(II) sulphate with deviating m, one has to weigh out m_x/m times the prescribed amount.
- The iron(II)-sulfosalicylic acid complex has a red colour, whereas sulfosalicylic acid is colourless; the colour changes, however, from red to yellow because the Fe-EDTA complex is coloured yellow.
- Boiling of $(NH_4)_2Fe(SO_4)_2 \cdot xH_2O$ in an acid medium is necessary to get complete dissolution of the small amount of Fe^{3+} which is always present.
- When approaching pH 2.5, the solution shows a more and more intense yellow colour due to hydrolysis of Fe^{3+}.
- Ammonium persulfate can also be used for oxidation.

APPENDIX A-20. STANDARDIZATION OF $MgSO_4 \cdot xH_2O$ WITH EDTA

Purpose

Since $MgSO_4 \cdot 7H_2O$ may lose crystal water on standing, the magnesium content of this salt should be determined before it is used as a standard.

Principle

A solution of magnesium sulphate is titrated with EDTA at pH 10 with Eriochrome Black T as an indicator.

Reagents

Buffer solution pH 10. Dissolve 54 g of ammonium chloride, NH_4Cl, in 350 ml of concentrated aqueous ammonia (25%) and dilute with water to 1000 ml.

EDTA solution, 0.05 mol/l. Standardized according to App. A-14.

Indicator solution. Eriochrome Black T, prepared according to App. A-5.

Procedure

Weigh out precisely about 0.37 g of $MgSO_4 \cdot xH_2O$ in a 300-ml erlenmeyer flask. Dissolve it in about 100 ml of water. Heat to 70–80 °C, add 10 ml of the buffer solution and 0.4 ml of the indicator solution. Titrate from wine-red to a blue end point with standardized 0.05 M EDTA solution.

Perform also a blank determination.

Calculation

The actual molecular weight (m_x) of the used magnesium sulphate is:

$$m_x = \frac{w}{t \times (a - b)}$$

in which

w = weight of $MgSO_4 \cdot xH_2O$, in mg;

t = titer of the EDTA solution, in mol/l;

a = volume of EDTA solution used for the analyte, in ml;

b = volume of EDTA solution used for the blank, in ml.

Remarks:

- If x = 7, then m = 246.48; when preparing standard solutions of magnesium, using magnesium sulphate with deviating m, one has to weigh out m_x/m times the prescribed amount.
- The buffer solution used for the standardization of EDTA (App. A-14) may be used as well.

APPENDIX A-21. STANDARDIZATION OF $KMnO_4$ WITH EDTA (AFTER REDUCTION)

Purpose

Since $KMnO_4$ may decompose on standing by influence of light, the manganese content of this salt should be determined before it is used as a standard.

Principle

A known amount of $KMnO_4$ is reduced to a Mn^{2+}/Mn^{3+} mixture by sodium sulphite. Hydroxylamine-HCl is added to reduce the manganese completely to Mn^{2+}. The solution is titrated with EDTA at pH 10 with Eriochrome Black T as an indicator.

Reagents

Sulphuric acid, $c(H_2SO_4)$ = 0.7 mol/l. Add carefully, while swirling, 40 ml of concentrated sulphuric acid (96%) to about 400 ml water. Allow to cool and make up to 1 litre.

Sulphite solution. Dissolve 5 g of sodium sulphite, Na_2SO_3, in 100 ml water.

Hydroxylamine-HCl solution. Dissolve 1 g of hydroxylamine-HCl, $NH_2OH \cdot HCl$, in 100 ml water.

Ammonia solution, $c(NH_3)$ = 4 mol/l. Dilute 75 ml of concentrated aqueous ammonia (25%) with water to 250 ml.

Tartrate solution pH 10. Dissolve 3.75 g of tartaric acid, $C_4H_6O_6$, in 35 ml of 4 M ammonia solution and dilute with water to 250 ml.

EDTA solution, 0.05 mol/l. Standardized according to App. A-14.

Indicator solution. Eriochrome Black T, prepared according to App. A-5.

Procedure

Weigh out precisely about 0.24 g of $KMnO_4$ in a 300-ml erlenmeyer flask. Dissolve it in about 50 ml of water. Add 7.5 ml of 0.7 M sulphuric acid and 11 ml of the Na_2SO_3 solution. Swirl until the solution is clear and colourless; if the solution is not yet colourless, add a few drops extra of the Na_2SO_3 solution. Add 10 ml of the hydroxylamine-HCl solution, 30 ml of the tartrate solution and 0.4 ml of the indicator solution. Titrate from red to a blue end point with standardized 0.05 M EDTA solution.

Perform also a blank determination.

Calculation

The actual molecular weight (m_x) of the used potassium permanganate is:

$$m_x = \frac{w}{t \times (a - b)}$$

in which

w = weight of $KMnO_4$, in mg;

t = titer of the EDTA solution, in mol/l;

a = volume of EDTA solution used for the analyte, in ml;

b = volume of EDTA solution used for the blank, in ml.

Remarks:

- Without decomposition, m = 158.03; when preparing standard solutions of manganese, using potassium permanganate with deviating m, one has to weigh out m_x/m times the prescribed amount.
- $KMnO_4$ can also be standardized with sodium oxalate (see App. A-12).

APPENDIX A-22. STANDARDIZATION OF $Ni(NO_3)_2 \cdot xH_2O$ WITH EDTA

Purpose

Since $Ni(NO_3)_2 \cdot 6H_2O$ may absorb water on standing, the nickel content of this salt should be determined before it is used as a standard.

Principle

A solution of nickel nitrate is titrated with EDTA at pH 10 with murexide as an indicator.

Reagents

Buffer solution pH 10. Dissolve 54 g of ammonium chloride, NH_4Cl, in 350 ml of concentrated aqueous ammonia (25%) and dilute with water to 1000 ml.

EDTA solution, 0.05 mol/l. Standardized according to App. A-14.

Murexide indicator. Suspend 50 mg of murexide in 5 ml of water to get a saturated solution. Filter. Prepare a fresh solution each day.

Procedure

Weigh out precisely about 0.44 g of $Ni(NO_3)_2 \cdot xH_2O$ in a 300-ml erlenmeyer flask. Dissolve it in about 100 ml of water. Add 10 ml of the buffer solution and 3 drops of the murexide indicator. Titrate from yellow to a violet end point with standardized 0.05 M EDTA solution.

Calculation

The actual molecular weight (m_x) of the used nickel nitrate is:

$$m_x = \frac{w}{t \times a}$$

in which

w = weight of $Ni(NO_3)_2 \cdot xH_2O$, in mg;

t = titer of the EDTA solution, in mol/l;

a = volume of EDTA solution used, in ml.

Remarks:

- If x = 6, then m = 290.79; when preparing standard solutions of nickel, using nickel nitrate with deviating m, one has to weigh out m_x/m times the prescribed amount.
- The buffer solution used for the standardization of EDTA (App. A-14) may be used as well.
- Nickel is a poisonous element. Work in a fume hood and make use of pipette fillers.
- Ni-murexide has a yellow colour. However, in the beginning of the titration the colour is yellow-green, which is due to a mixture of Ni-EDTA and Ni-murexide. At the endpoint the solution becomes deep blue, caused by the mixed colours of (blue) Ni-EDTA and (purple) murexide.

APPENDIX A-23. STANDARDIZATION OF $Pb(NO_3)_2$ WITH EDTA

Purpose

Since $Pb(NO_3)_2$ may absorb water on standing, the lead content of this salt should be determined before it is used as a standard.

Principle

A solution of lead nitrate is titrated with EDTA at pH 10 in the presence of tartrate ions with Eriochrome Black T as an indicator.

Reagents

Ammonia solution, $c(NH_3)$ = 4 mol/l. Dilute 75 ml of concentrated aqueous ammonia (25%) with water to 250 ml.

Tartrate solution pH 10. Dissolve 3.75 g of tartaric acid, $C_4H_6O_6$, in 35 ml of 4 M ammonia solution and dilute with water to 250 ml.

EDTA solution, 0.05 mol/l. Standardized according to App. A-14.

Indicator solution. Eriochrome Black T, prepared according to App. A-5.

Procedure

Weigh out precisely about 0.50 g of $Pb(NO_3)_2$ in a 300-ml erlenmeyer flask. Dissolve it in about 100 ml of water. Add 20 ml of the tartrate solution. Swirl until the appearing precipitate has dissolved. Then add 0.2 ml of the indicator solution and titrate from red-violet to a blue end point with standardized 0.05 M EDTA solution.

Perform also a blank determination.

Calculation

The actual molecular weight (m_x) of the used lead nitrate is:

$$m_x = \frac{w}{t \times (a - b)}$$

in which

w = weight of $Pb(NO_3)_2$, in mg;
t = titer of the EDTA solution, in mol/l;
a = volume of EDTA solution used for the analyte, in ml.
b = volume of EDTA solution used for the blank, in ml.

Remarks:

- When no water is absorbed (i.e., x = 0), then m = 331.21; when preparing standard solutions of lead, using lead nitrate with deviating m, one has to weigh out m_x/m times the prescribed amount.
- Lead is a very poisonous element. Work in a fume hood and make use of pipette fillers.
- The tartrate solution is added to prevent precipitation of Pb as a hydroxide.

APPENDIX A-24. STANDARDIZATION OF $ZnSO_4 \cdot xH_2O$ WITH EDTA

Purpose

Since $ZnSO_4 \cdot 7H_2O$ may lose crystal water on standing, the zinc content of this salt should be determined before it is used as a standard.

Principle

A solution of zinc sulphate is titrated with EDTA at pH 10 with Eriochrome Black T as an indicator.

Reagents

Buffer solution pH 10. Dissolve 54 g of ammonium chloride, NH_4Cl, in 350 ml of concentrated aqueous ammonia (25%) and dilute with water to 1000 ml.

EDTA solution, 0.05 mol/l. Standardized according to App. A-14.

Indicator solution. Eriochrome Black T, prepared according to App. A-5.

Procedure

Weigh out precisely about 0.43 g of $ZnSO_4 \cdot xH_2O$ in a 300-ml erlenmeyer flask. Dissolve it in about 100 ml of water. Add 10 ml of the buffer solution and 0.4 ml of the indicator solution. Titrate from wine-red to a blue end point with standardized 0.05 M EDTA solution.

Perform also a blank determination.

Calculation

The actual molecular weight (m_x) of the used zinc sulphate is:

$$m_x = \frac{w}{t \times (a - b)}$$

in which

w = weight of $ZnSO_4 \cdot xH_2O$, in mg;

t = titer of the EDTA solution, in mol/l;

a = volume of EDTA solution used for the analyte, in ml;

b = volume of EDTA solution used for the blank, in ml.

Remarks:

- If $x = 7$, then $m = 287.56$; when preparing standard solutions of zinc, using zinc sulphate with deviating m, one has to weigh out m_x/m times the prescribed amount.
- The buffer solution used for the standardization of EDTA (App. A-14) may be used as well.

APPENDIX A-25. PREPARATION OF AZOMETHINE-H (REAGENT FOR BORON)

Purpose

Azomethine-H can normally be purchased commercially. If this is for any reason not possible, the reagent can be made in the laboratory.

Principle

In acid medium, salicylaldehyde reacts with an aminonaphtoldisulphonic acid to give Azomethine-H.

Reagents

Concentrated hydrochloric acid, 36%, c(HCl) = 12 mol/l, d = 1.19 g/ml.

Sodium hydroxide solution, c(NaOH) = 2.5 mol/l. Dissolve 10 g of sodium hydroxide, NaOH, in some water. Allow to cool and make up to 100 ml.

Na-AND: 8-amino-1-naphtol-3,6-disulphonic acid, monosodium salt (= 4-amino-5-hydroxy-2,7-naphtalene-disulphonic acid, monosodium salt, $C_{10}H_8NO_7S_2Na$).

Salicylaldehyde, (2-hydroxybenzaldehyde, $C_7H_6O_2$).

Ethanol, 96%.

Procedure

Weigh out 10 g of Na-AND in a 1-litre polythene beaker. Add 500 ml water and dissolve the Na-AND. Neutralize to pH 7 by dropwise addition of sodium hydroxide solution 2.5 M. Then add concentrated hydrochloric acid until pH 1.5 is reached; a precipitate will be formed. Heat to 50–70 °C until all precipitate has dissolved. Add 10 ml of salicylaldehyde and stir vigorously during 1 h. Allow to stand overnight.

Decant the supernatant liquid and collect the yellow precipitate of Azomethine-H in centrifuge tubes. Centrifuge for 10 min at 3600 **g** (angle rotor) and decant the supernatant. Add ethanol to the sediment and bring it into suspension. Centrifuge again for 10 min at 3600 **g** and discard the supernatant. Repeat this treatment another 3 times.

Transfer the precipitate with ethanol to a Büchner funnel and remove the ethanol as completely as possible by suction. Dry the Azomethine-H at 100–105 °C in a preheated oven with *power switched off*. Store in a desiccator above silicagel.

Remarks:

- Never use ultrasonic energy to bring the Azomethine-H into suspension, because this will desintegrate the compound.

- While drying the Azomethine-H the oven must not switch on or off, because there may be some explosion danger.
- According to our experience, home-made Azomethine-H gives lower zero standard values, because of less own colouration.
- According to our experience, the use of KOH leads to higher blank values than NaOH.

APPENDIX A-26. RECRYSTALLIZATION OF BORAX

Purpose

Borax may easily lose crystal water, which renders the salt useless as a standard substance. The desired decahydrate can readily be obtained by recrystallization.

Principle

Sodium borate decahydrate is recovered by crystallization from a saturated solution at room temperature.

Reagents

Borax (sodium tetraborate, $Na_2B_4O_7 \cdot xH_2O$).
Ethanol, 96%.
Diethyl ether, analytical grade.

Procedure

Dissolve 30 g of borax in 100 ml water by gentle heating. Cool to room temperature. Filter the newly formed crystals on a Büchner funnel by suction. Wash twice with 15-ml portions of water, then twice with 15-ml portions of ethanol and finally twice with 15-ml portions of diethyl ether. Each washing must be followed by suction to remove the wash liquid. After the final washing, the solid is spread in a thin layer on a watch glass and allowed to stand at room temperature for 12–18 h. Store in a well-stoppered bottle.

Remarks:

- Care must be taken that the crystallization does not take place above 55 °C; above this temperature there is a possibility of the formation of the pentahydrate.
- Stored in a well-stoppered bottle, the recrystallized borax can still be used after 3–4 weeks without noticeable change.
- Diethyl ether is extremely flammable. Work in a well-ventilated fume hood; extinguish any flames, do not smoke, and avoid sparks from electrical apparatus switching on or off.

REFERENCES

J. Basset et al. Vogel's textbook of quantitative inorganic analysis, 4th ed., Longman, London (1978) p. 300.

APPENDIX A-27. DETERMINATION OF CRYSTAL WATER CONTENT IN BORAX

Purpose

Since borax, $Na_2B_4O_7 \cdot 10H_2O$, may easily lose crystal water on standing, the actual water content of this salt should be determined before it is used as a standard.

Principle

A solution of borax is titrated with acid with methyl red-bromocresol green as an indicator.

Reagents

Hydrochloric acid, c(HCl) = 0.1 mol/l. Standardized according to App. A-9.

Indicator solution. Mixed indicator methyl red-bromocresol green, prepared according to App. A-5.

Procedure

Weigh out precisely about 600 mg of the suspected borax, $Na_2B_4O_7 \cdot xH_2O$, in a 300-ml erlenmeyer flask. Dissolve it in about 40 ml of distilled water (CO_2-free), and add 3 drops of the indicator solution. Titrate with standardized 0.1 M hydrochloric acid until the green colour turns via grey into rose.

Calculation

The actual molecular weight (m_x) of the used borax is

$$m_x = \frac{w \times 2}{t \times a}$$

in which
w = weight of $Na_2B_4O_7.xH_2O$, in mg;
t = titer of the acid, in mol/l;
a = volume of acid consumed, in ml.

Remarks:

- If $x = 0$, then $m = 201.22$; the difference between m_x and m, divided by 18, gives the number of crystal water molecules actually present in a molecule of this lot of borax.
- The calculation of the amount of borax to be weighed out is mostly based on the use of $Na_2B_4O_7 \cdot 10H_2O$ ($m_{10} = 381.37$). In that case, when preparing standard solutions of boron using borax with deviating m, one has to weigh out m_x/m_{10} times the prescribed amount. Alternatively, the borax may be recrystallized (see App. A-26).
- The titration may be done also with sulphuric acid, but the colour change is less sharp then.

MIX
Papier aus verantwortungsvollen Quellen
Paper from responsible sources
FSC® C105338

If you have any concerns about our products,
you can contact us on
ProductSafety@springernature.com

In case Publisher is established outside the EU,
the EU authorized representative is:
Springer Nature Customer Service Center GmbH
Europaplatz 3, 69115 Heidelberg, Germany

Printed by Libri Plureos GmbH
in Hamburg, Germany